SciencePlus

Interactive Explorations

CD-ROM for Macintosh® and Windows®

Teacher's Guide

Level Blue

HOLT, RINEHART AND WINSTON
Harcourt Brace & Company

Austin • New York • Orlando • Atlanta • San Francisco • Boston • Dallas • Toronto • London

TO THE TEACHER

Imagine having access to a fully equipped laboratory where your students could study questions and problems related to DNA fingerprinting, the particle theory of matter, electricity and magnetism, and the Coriolis effect. This is exactly what is possible with *SciencePlus Interactive Explorations*. By using this innovative CD-ROM program, students get valuable experience in a unique laboratory setting.

Welcome to Dr. Crystal Labcoat's laboratory, where with the click of a mouse, students will have access to a variety of scientific tools and equipment and will be challenged to solve some perplexing problems and mysteries. Dr. Labcoat operates a virtual laboratory, and your students are her lab assistants. Under her guidance, your students will perform some amazing and highly interesting scientific experiments and studies. But be prepared—although Dr. Labcoat provides the lab and the equipment, your students provide the brainpower.

This *Teacher's Guide* consists of the following components:

- **User's Guide**
  The User's Guide provides important technical information about the program, including its installation, features, and use.

- **Teaching Notes, Worksheets, and Handouts**
  Organized by exploration, this information includes background material and worksheets that guide students through the CD-ROM experience and allow them to record their answers on paper, rather than electronically. In addition, the Computer Database articles (also called CD-ROM articles) are provided so that they can be used as handouts. The worksheets and handouts make the program more flexible in cooperative groups or when computer time is limited.

- **Answer Keys**
  Worksheet pages with overprinted answers are provided for each exploration to make grading worksheets and fax forms fast and efficient.

Photo/Art Credits

Abbreviations used: (b) bottom, (l) left.

All work, unless otherwise noted, contributed by Holt, Rinehart and Winston.

Front Cover: (bl) Jeff Foot; Marty Cornado.

All photos used in the *SciencePlus Interactive Explorations CD-ROM*, Level Blue, contributed by Holt, Rinehart and Winston, unless otherwise noted below:

Exploration 2 (USS Birmingham submarine) courtesy of U.S. Naval Institute.

Exploration 4 (Antarctica map) Mountain High Maps® Copyright © 1995 Digital Wisdom, Inc.

SCIENCEPLUS is a registered trademark licensed to Holt, Rinehart and Winston.

Printed in the United States of America

ISBN 0-03-018542-4   3 4 5 6 7 8 9 021 00 99 98

CONTENTS

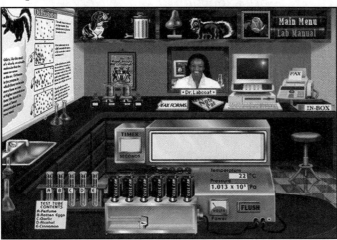

Ms. Dee Foushen needs help sniffing out the solution to a problem. She wants to know the best chemical to use for an odor alarm that would warn her hearing- and sight-impaired students in the event of a fire.

Ms. Diane Sittie has come up with a bright idea: an under-water lamp for scuba divers that neither sinks nor floats. Unfortunately, she is in the dark about some of the details of her invention and needs help keeping her idea afloat.

Mr. Norm N. Cline is fighting an uphill battle in the design of an amusement park ride that consists of a slide and toboggans. He needs help selecting the best materials and toboggan size to avoid unnecessary friction with park patrons.

Research scientists on the Mertz Glacier in Antarctica are expecting an air-delivery of fresh supplies. Captain Corey O. Lease must confirm her flight plan so that she can straighten up and fly right to the research site.

Seymore Rhodes wants to create a lighted bicycle helmet that will make riding his bike to school safer. Unfortunately, he's not an electrical genius—he needs help seeing the light at the end of the tunnel.

Exploration 6: Sound Bite!

A loud humming noise from a new ice cream shop has some guinea pigs at the neighboring pet store in an uproar. Mr. Cy Lintz wants to know how to use active sound control to give his guinea pigs some peace and quiet.

Exploration 7: In the Spotlight

Ms. Iris Kones' first production at a community theater is really in the spotlight. She needs help filtering through her lighting options to determine how many different colors of light she can produce.

Exploration 8: DNA Pawprints

Ms. Jean Poole, a local dog breeder, is afraid she's barking up the wrong tree when it comes to completing pedigrees. She needs help figuring out which of her male dogs sired her younger dogs so that she can enter the pups in a dog show.

USER'S GUIDE

SYSTEM REQUIREMENTS

Before you begin using *SciencePlus Interactive Explorations,* you will need to acquire the necessary equipment and set it up properly. The complete setup includes a computer (IBM®-compatible or Macintosh®-compatible) connected to a CD-ROM drive. Audio headphones are optional. To use the program with a network, you will also need additional cables to connect the machines, a dedicated network file server, and a network operating system.

A note concerning minimum requirements: *Although the program will run on machines with the minimum requirements listed below, we strongly recommend that the program be used on newer model computers (040 Macintoshes and 486 PCs or higher) that are more efficient at handling the demands of multimedia. If you run the program on lower-end machines, you may experience slow response times to clicks, longer loading times for video and sound, and slower animations, as well as occasional dropped video frames and sound.*

COMPUTERS

Macintosh®-Compatible Computers
Minimum Requirements: • 68030 CPU running at 25 MHz or higher **(Highly recommended: 68040 CPU running at 20 MHz or higher)** • 13-inch or larger color monitor capable of displaying 256 colors at 640 × 480 resolution • System 7.1 or higher • Double-speed or higher CD-ROM drive • 8 MB of RAM • 30–40 MB of free memory on hard drive if you plan to install individual explorations (explorations can also be run directly from the CD-ROM) • QuickTime® for Macintosh® (provided with the program) • Internal/external speaker(s); headphones (recommended for classroom settings)

IBM®-Compatible Computers
Minimum Requirements: • 80386 DX running at 25 MHz or higher **(Highly recommended: 80486 running at 33 MHz or higher)** • 13-inch or larger color monitor capable of displaying 256 colors at 640 × 480 resolution • Windows® 3.1 or higher • Double-speed or higher CD-ROM drive • 8 MB of RAM • 30–40 MB of free memory on hard drive if you plan to install individual explorations (explorations can also be run directly from the CD-ROM) • QuickTime® for Windows® (provided with the program) • Sound Blaster™ or other compatible sound card • Internal/external speaker(s); headphones (recommended for classroom settings)

PRINTERS

Minimum Requirements:
Minimum Requirements: • Laser printer, ink-jet printer, or 24-pin dot-matrix printer

vi   *SCIENCEPLUS* INTERACTIVE EXPLORATIONS TEACHER'S GUIDE • LEVEL BLUE

HRW material copyrighted under notice appearing earlier in this work.

. . . ON MACINTOSH®-COMPATIBLE COMPUTERS

1. Place the CD-ROM in the CD-ROM drive.

2. A window will appear with the program's "Read Me" file and the **SciPlus Level Blue Installer** icon. Double-click this icon and follow the procedures on the screen.

3. For information concerning installation of individual explorations, please see the "Read Me" document on this screen.

4. After installation is complete, a **SciencePlus Blue** folder will appear on your hard drive. Open the folder and double-click the **SciencePlus Blue** icon to launch the program.

. . . ON IBM®-COMPATIBLE COMPUTERS (WITH WINDOWS®)

1. Place the CD-ROM in the CD-ROM drive.

2. Locate the **Install** and double-click it.

3. Once the program is installed, a window will appear with the program's "Read Me" file, the **SciPlus Blue** icon, and the **Uninstall SP Blue** icon.

4. Launch the program by clicking the **SciPlus Blue** icon found in the directory or folder of Program Manager.

Note: *For details on how to do a custom installation of individual explorations to improve performance, please see the "Read Me" file located on the CD-ROM.*

Logging on to *SciencePlus Interactive Explorations* is quick and simple. After launching the program, a log-on display will ask your students if they are working as guests, as individuals, or as a group.

. . . AS AN INDIVIDUAL

To log on as an individual, a student follows this procedure:

1. The student clicks the **Individual** button. A dialog box will appear asking the student to type in his or her first and last name, and the teacher's name and class or period.
2. The student clicks **Enter** or presses **Return.**
3. When the main menu appears, the student chooses an exploration.

. . . AS A GROUP

To log on as a group, students follow this procedure:

1. Students click the **Group** button. Students are then asked to type in their one-word group name, the teacher's name, the class or period, and the names of the members of their group.
2. Students click **Enter** or press **Return.**
3. When the main menu appears, the group chooses an exploration.

. . . AS A GUEST (Nonassessed Use of the Program)

The log-on contains a guest feature that allows you or your students to do an exploration without engaging the assessment function of the program. In other words, what is completed using the guest feature will not be graded. To log on as a guest, a student clicks the **Guest** button. When the main menu appears, the student chooses an exploration.

An Important Note: *Please be sure to consult the "Read Me" file found on the CD-ROM. There you will find important, up-to-date information concerning general technical issues and changes that may not be present in this guide. Likewise, an additional "Read Me" file is contained in the Assessment Tools folder of the CD-ROM. This file identifies any updates for using the assessment tools.*

The Main Menu allows you to select the following items:

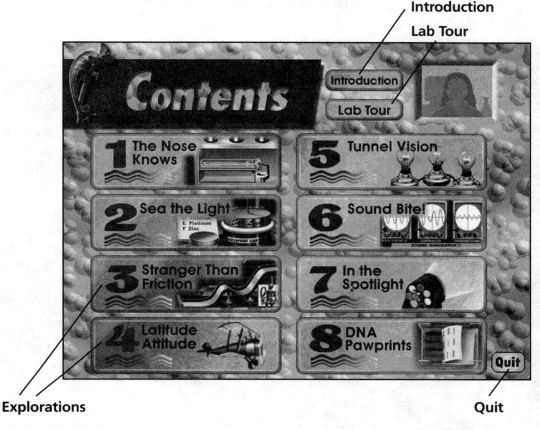

Introduction

Lab Tour

Explorations

Quit

• Introduction

Click the **Introduction** button to get a general overview of the explorations.

• Lab Tour

Click the **Lab Tour** button to get a quick yet comprehensive tour of Dr. Crystal Labcoat's laboratory. This is an excellent way for both you and your students to get acquainted with the standard equipment and features of the lab. Of course, depending on the exploration and the problem to be solved, Dr. Labcoat has a wide variety of specialized equipment, which is described in the exploration in which the equipment is used.

• Interactive Explorations

Simply click any one of the eight **Exploration** buttons to start that particular exploration. Should you want a quick overview of an exploration, move the cursor over any exploration button, and a brief description of the exploration will automatically appear on the screen.

• Quit

You can exit the program by clicking the **Quit** button on the main menu. Within an exploration, you can also end your session by selecting **Quit** in the pull-down menu under **File.**

Dr. Labcoat's laboratory is a rich and functional scientific setting where students can practice their scientific problem-solving skills as well as their process skills as they try to solve a variety of science-related problems and mysteries. The following information will help you navigate and use the features of this unique laboratory.

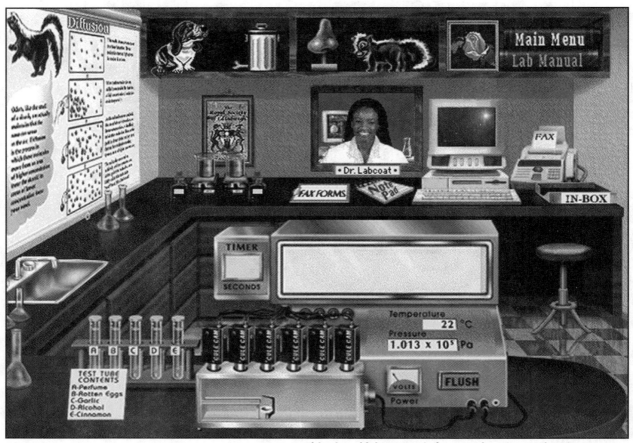

This virtual laboratory is from Exploration 1, The Nose Knows.

NAVIGATION

Navigation is accomplished by moving the mouse, which operates the cursor. You will notice that four types of cursors are used in the explorations.

Arrow Cursor		This point-and-click cursor is used to close pop-up windows as well as to set variables in experiments that require adjustments before a simulation can begin.
Pointing-Finger Cursor		This point-and-click cursor indicates the areas of the lab that are active. Activate a selection by clicking it.
Hand Cursor		This cursor indicates movable objects. To move objects, click and hold the mouse. You will be able to drag objects to lab equipment for testing as well as for classification and storage.
Bar Cursor		This point-and-click cursor appears in fields that require typing or word processing.

TOOL BAR

A tool bar featuring pull-down menus will display the following items:

File	Edit	Sound	Windows	Options
Quit	Copy Select All **Active when you are copying text from the Computer Database**	Level 0 Level 1 Level 2 Level 3 Level 4 Level 5 Level 6 Level 7	Notepad **Access to other pop-up variable panels available in certain explorations**	English Audio Spanish Audio **Available for the Lab Manual only**

LAB MANUAL

The Lab Manual provides a short explanation of the purpose and the operation of each piece of equipment in the lab. This is a handy reference for students who may need additional help.

To page through the Lab Manual, click the tabs at the lower right corner of the page. You can play audio instructions in English by clicking the megaphone-shaped icon. To hear instructions in Spanish, click **Options** on the tool bar and then select **Spanish Audio.** To hide the Lab Manual, click the **Close** box in the upper right corner.

FAX MACHINE

The Fax Machine is a primary means of communication both to and from the lab. Incoming fax messages identify problems to be solved. Outgoing fax messages are generated by students as they solve the problems and are asked to communicate their findings. Faxes often contain many pages. To page through faxes, click the tabs at the lower right corner of the page. To hide faxes, click the **Close** box in the upper right corner.

IN-BOX

The In-Box is where fax messages and other correspondence are kept for reference at any time during the exploration. To page through faxes and other correspondence, click the tabs at the lower right corner of the page. To hide the correspondence, click the **Close** box in the upper right corner.

NOTEPAD

The Notepad is always available for jotting down notes and observations. Simply click the Notepad and start taking notes. Students can even paste articles from the Computer Database into the Notepad. Since the contents of the Notepad are not saved by the program, students should print their Notepads before leaving an exploration if they are interested in keeping a record of their notes.

To paste text from the Computer Database into the Notepad, simply highlight the desired passages (or click **Select All**) and then select **Copy** from the pulldown **Edit** menu. Go to the Notepad and click the **Paste** icon. The text that you copied from

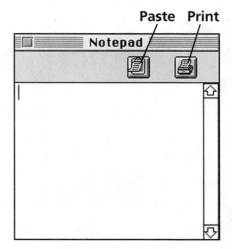

Paste Print

the Computer Database will appear in the Notepad. **To print the contents** of the Notepad, simply click the **Print** icon.

More advanced students may choose to take notes or paste information from the Computer Database into other documents (such as those created by SimpleText) by running a separate word-processing application. Although this method is more complex, it allows students to save their notes electronically. For information concerning the simultaneous use of two applications, refer to the user's guide that accompanies your system's software.

COMPUTER DATABASE

Students can use this Computer Database to easily access information on a variety of subjects. The information is organized into articles that contain text as well as illustrations, photographs, and video. The Computer Database contains information that is vital to solving the problem or mystery.

To access articles that are relevant to an exploration at hand, students simply scroll to the applicable topic and click on subtopics to view that information. Students can also access the complete database of articles by clicking the **Database Index** button. To return to the exploration from which they accessed the database, students click **Table of Contents.**

It is important to note that students can copy text from the Computer Database into their notepads or other word-processing documents for printing. Images and video appearing in the Computer Database, however, cannot be pasted into the Notepad.

FAX FORMS

Fax forms located on the clipboard contain the forms necessary to create a new fax. Sending a fax is how students will communicate their solutions to the problems. Once a fax form has been completed, it is sent by clicking the **Send It** button. An appropriate response will be forthcoming from both the requester of the information and from Dr. Labcoat.

CALCULATOR

In explorations that require mathematical calculations, a calculator is provided in the lab. Simply click the calculator to bring up the calculator's keypad.

ASSESSMENT TOOLS

A variety of assessment tools are available to make grading and record keeping as simple as possible. The assessment tools are located in the following folders, depending on which CD-ROM level (Green, Red, or Blue) you are using.

SciencePlus Level	Folder Name
Level Green	**adminSPG**
Level Red	**adminSPR**
Level Blue	**adminSPB**

If you are using a Macintosh®, you can access the Assessment Tools folder by entering the Preferences folder of your System folder. **If you are using a PC,** you will find the Assessment Tools folder in the Windows® folder. You can also do a search for the Assessment Tools folder by entering the folder name as identified in the chart above.

When you open an Assessment Tools folder, you will find the following folders:

- Read Me
- Student Reports
- Answer Keys
- File Management

To view the contents of any folder, simply double-click the folder.

READ ME FOLDER
All updates to the assessment tools component of the program can be found in this printable file.

STUDENT REPORTS FOLDER
You can access student reports after each class session or at the end of the day by double-clicking the folder labeled "Student Reports." In the folder, you will see your students' work in the form of SimpleText files. These documents can be opened in SimpleText or in another word-processing application, if you choose.

Each student file or record has a prefix consisting of up to five letters. In the case of individual students, the prefix consists of the first four letters of the student's last name and the first initial of the student's first name; in the case of groups, the group's name (up to five letters) will appear. The prefix is followed by a period (.) and a three-letter suffix (indicating the level of *SciencePlus Interactive Explorations* and the exploration number).

Consider the following examples:

File Name	Individual or Group Name	Level	Exploration
Smitj.GR1	Smith, Jennifer	Green	1
Smitp.GR2	Smith, Peter	Green	2
Aces.RD7	Aces	Red	7
Diamo.BL8	Diamonds	Blue	8

All student records relate directly to the fax forms of the *SciencePlus Interactive Explorations,* where students produce their work. Each student report contains the following sections:

Student/Class Information Section
This section contains:
- Student name or group name with names of group members
- Teacher's name
- Class or period
- The date the exploration was conducted
- The duration of the student's work within the exploration

Computer-Graded Section
This section contains the student's responses to the close-ended questions on the fax form. These questions, which are indicated with an asterisk (*), will have already been graded by the computer. Fifty (50) points are possible.

Teacher-Graded Section
This section consists of the answers to the open-ended questions on the fax form. This section must be read and evaluated by the teacher. Fifty (50) points are possible.

Scoring Section
This section consists of three scores: the computer-graded score, the teacher-graded score, and the composite score (final grade). Suggestions for grading student work can be found in the Answer Keys folder.

Teacher Comments Section
This section provides a space for making comments on your students' work. Simply type in your comments as you view the Student Report, or you can print out the Student Report and write in your comments manually.

ANSWER KEYS FOLDER
The Answer Keys folder contains eight answer keys, which correspond to the eight explorations for a particular level. The files are named as indicated in the chart below. Please note that this chart only shows the file names for Exploration 1 in levels Green, Red, and Blue. For Exploration 2, the file names would end in "2."

File Name	Level	Exploration
AnkeyGR1	Green	1
AnkeyRD1	Red	1
AnkeyBL1	Blue	1

If desired, you can customize the names of these files simply by renaming them. Also, if you have installed Level Green or Red on the same machine, their respective contents will be found in another folder on your hard drive.

To use an answer key:

1. Double-click the answer key you wish to view.

2. Resize the answer key if you would like to view it on-screen along with a student report.

3. Grade your students' responses to the open-ended questions on the student report. Remember, the computer-graded questions are scored automatically.

4. If desired, print out hard copies of the answer key, the student's report, or both.

FILE MANAGEMENT FOLDER

The File Management folder contains an example of how you might set up class folders for storing your files. You can use this simple folder structure by changing the names of folders or adding more folders to suit your needs.

We suggest that you set up your records by class. At the end of each class, move student files into their corresponding class folder. You may decide to set up folders within each class folder by student or by exploration number. Viewing files by date and time will be helpful in determining the class folder into which a student file can be inserted.

If you need help structuring your folders, please consult your computer manual for more in-depth instructions.

NETWORKING STUDENT REPORTS

The student reports in the Assessment Tools folder can be networked. This allows you to gather student reports to a central computer. To do this, you will need a dedicated network file server meeting the following requirements:

Macintosh®-Compatible Computers

Minimum Requirements:
- 68040 CPU running at 33 MHz or higher, such as a Quadra or PowerPC
- 16 MB of RAM
- Recommended: 20 K per user per semester (or 40 K per user per year)

IBM®-Compatible Computers

Minimum Requirements:
- Minimum 486DX running at 33 MHz or higher
- 16 MB of RAM
- Recommended: 20 K per user per semester (or 40 K per user per year)

You will also need Ethernet cards and cables, Novell Netware® 3.0 or higher, Windows® NT ™ 3.51 or higher, or AppleShare® 3.0 or higher network operating system.

OPTIMIZING PERFORMANCE

There are many things you can do to optimize the performance of *SciencePlus Interactive Explorations*. Consider the following list of options:

1. Make sure that your monitor's resolution is set to 256 colors. Running the program on "Thousands of Colors" may slow the program down.

2. Do a custom installation of an exploration rather than running the program completely off the CD-ROM. The program will perform better if an exploration is installed onto your hard drive. Remove explorations from the hard drive when you move on to a new exploration.

3. If your hard drive is more than 80 percent full, performance may suffer. Remove old files and applications that are no longer pertinent or useful. Consider storing them on another hard drive or with alternative methods of data storage (such as Zip™ cartridges).

4. Use a utility to optimize your hard drive. Refer to the manual that came with your computer for more information about optimizing your hard drive.

5. If you are running other applications in the background—no matter how simple or complex the application—performance will suffer. Quit all applications other than *SciencePlus Interactive Explorations*.

6. If you are on a network or on the Internet, your chances of experiencing freezes, crashes, and poor video or audio are much greater than if you are not.

7. Make sure you have the most current version of QuickTime 2.1 and QuickTime Powerplug extensions and the latest version of Sound Manager extension (v. 3.2). Using older versions of these extensions may result in reduced video quality.

8. If you have been running any memory-intensive applications prior to running *SciencePlus Interactive Explorations,* restart your machine before running the program.

9. If the total memory requirement of your system software and this program approaches the limits of your machine's total RAM, performance will suffer. Consider upgrading your machine's capability by adding more RAM. Also consider using an extensions-management utility to turn off extensions not being used by the application or temporarily placing unused extensions in a folder labeled "Disabled Extensions."

10. Turn off your computer's Virtual Memory.

Suggestion: *If students are working in a setting that includes lower-end machines as well as newer models, you may want to rotate students or groups from one machine to another. Have your students do their research with the Computer Database section of the program on lower-end machines and do the main experiments using the better performing machines.*

TECHNICAL SUPPORT INFORMATION

At Holt, Rinehart and Winston we recognize the importance of providing you with the answers and help you need to use our quality instructional-technology products to their fullest potential.

Because systems, technology, and content are often inseparable, HRW has assembled a team of dedicated technical and teaching professionals and a suite of comprehensive support services to provide you with the support you deserve, 24 hours a day, 7 days a week.

Technical Support Line	800-323-9239

The HRW Technical Support Line, which operates from 7 A.M. to 6 P.M. Central Standard Time, Monday through Friday, puts you in touch with trained Support Analysts who can assist you with technical and instructional questions on all of HRW's instructional technology products.

Technical Support on the World Wide Web	http://www.hrwtechsupport.com

Contact the HRW Technical Support Center 7 days a week, 24 hours a day, at our site on the World Wide Web. Simply select the product you are interested in, and with a click of the mouse you can receive comprehensive solutions documents, answers to the most frequently asked questions, product specifications and technical requirements, and program updates from our FTP site. You can also contact our analysts at the Support Center using the following E-mail address: **tsc@hrwtechsupport.com**

Technical Support via Fax	800-352-1680

Get the solutions you need with the HRW Technical Support Center's fax-on-demand service. Simply give us a call at our toll-free number to receive product-specific solutions within minutes. Our fax-on-demand service is available 7 days a week, 24 hours a day.

The Nose Knows

Key Concepts

Diffusion is the process in which particles move from an area of higher concentration to an area of lower concentration. The sense of smell can be used to detect warning odors and to protect humans from danger.

Summary

Ms. Dee Foushen is the director of a school for the hearing- and sight-impaired. She needs help designing a special fire alarm to ensure the safety of her students. Since some of her students cannot see warning lights or hear a typical alarm, she wants to install an "odor alarm" that would release an odorous chemical in the building in the event of a fire. Ms. Foushen has sent five samples of odorous chemicals to Dr. Labcoat's lab and wants to know which one would be the best choice.

Mission

Choose the best chemical to use for an odor alarm.

Solution

Cinnamon is the best odorous chemical to use for an odor alarm because its scent diffuses fairly quickly through the air and it is not a dangerous substance. Rotten eggs and alcohol both diffuse more quickly than cinnamon does, but these chemicals could be dangerous to students if used for the odor alarm.

Background

Although humans rely primarily on sight and sound to gather information, the human sense of smell has some interesting functions. Because of the anatomical structure of the brain, the sense of smell is closely associated with memory. Information from other sensory neurons (sight, touch, and hearing) gets routed through the thalamus only; information from olfactory neurons (including those that stimulate taste) goes to the thalamus as well as to portions of the brain associated with memory, namely the hippocampus (short-term memory) and the amygdala (long-term memory). This makes the sense of smell a powerful memory stimulator. The connection between smell and memory explains why some odors immediately trigger memories of a person, place, or time in the past.

Scientists are exploring the connection between smell and memory in research for Alzheimer's disease. This neurological disorder causes interruptions in the transmission of nerve impulses across the synapses in the brain. Alzheimer's causes concentrated synaptic loss in the limbic system, where the hippocampus and amygdala are located. As a result, many people with progressive Alzheimer's suffer from memory loss as well as a diminished sensitivity to smell, a condition called anosmia. Because of the difficulty in diagnosing Alzheimer's, some scientists have suggested monitoring declines in sensitivity to smell as a way to help identify the early stages of Alzheimer's.

Teaching Strategies

Students may be tempted to choose a substance for Ms. Foushen's odor alarm based only on its rate of diffusion. Emphasize that this is not the best approach, as it may lead students to choose an unsafe substance. For example, rotten eggs contain hydrogen sulfide, a toxic chemical, and alcohol is flammable. Encourage students to research each substance in the CD-ROM articles before making a decision.

As an extension of this Exploration, you may want to conduct a classroom activity such as the following: Set up an odor kit of various unknown odors (available through biological supply companies), and have students time how long it takes to sense each odor as the smell particles diffuse through the air. Students could then recommend an odor for use in an odor alarm. You can also ask students about ways to increase the rate of diffusion of each odor. For example, substances diffuse faster at higher temperatures.

You may also wish to discuss with students the possible limitations of implementing an odor alarm in Ms. Foushen's school for the hearing- and sight-impaired. Questions to ask students might consist of the following: What kind of sensor would detect the fire? How would the cinnamon be stored and released? Would the cinnamon be messy? Would it stain clothing or cause allergic reactions? If someone brought a fresh-baked cinnamon roll into the building, would it be mistaken for the odor alarm? Can you think of any other ways to notify the students of an unsafe situation?

Bibliography for Teachers

Taubes, Gary. "The Electronic Nose." *Discover,* 17 (9): September 1996, p. 40.

Trum Hunter, Beatrice. "The Sales Appeal of Scents." *Consumers' Research Magazine,* 78 (10): April 1995, p. 48.

Bibliography for Students

Lipkin, Richard. "Tracking an Undersea Scent." *Science News,* 147 (5): February 4, 1995, p. 78.

Schwenk, Kurt. "The Serpent's Tongue." *Natural History,* 104 (4): April 1995, p. 48.

Other Media

Diffusion and Osmosis
  Videotape
  Encyclopædia Britannica Educational Corporation
  310 S. Michigan Ave.
  Chicago, IL 60604-9839
  800-554-9862

In addition to the above video, students may find relevant information about diffusion by exploring the Internet. Interested students can search for articles using keywords such as *diffusion, osmosis,* and *semipermeable membranes.* Students can also access information about the *sense of smell* by exploring the Internet.

The Nose Knows

1. Ms. Foushen needs your help sniffing out the solution to a problem. What has she asked you to do?

_____

_____

_____

_____

2. Explain the process of diffusion. (Hint: Check out the wall chart in the lab.)

_____

_____

_____

_____

_____

_____

3. What is the difference between diffusion and osmosis? (If you're not sure, check out the CD-ROM articles.)

_____

_____

_____

_____

_____

4. What is the equipment on the front table in Dr. Labcoat's lab designed to do?

_____

_____

_____

_____

_____

5. Use the equipment to conduct the necessary tests, and record your data in the table below.

Test tube	Test-tube contents	Time to diffuse (sec.)
A	perfume	
B	rotten eggs	
C	garlic	
D	alcohol	
E	cinnamon	

6. Why are the temperature and pressure kept constant for this experiment? (If you're not sure, check out the CD-ROM articles.)

_____

_____

_____

_____

_____

7. What does the equipment on the back counter in Dr. Labcoat's lab demonstrate?

_____

_____

_____

_____

8. How do you smell an odorous chemical? Use the CD-ROM articles to help you explain how your sense of smell works.

_____

_____

_____

_____

_____

_____

Record your answers in the fax to Ms. Foushen.

FAX

To:	Ms. Dee Foushen (FAX 512-555-7003)
From:	
Date:	
Subject:	The Nose Knows

Which of the five samples do you recommend that I use for the fire alarm?

Alcohol	Cinnamon	Garlic	Perfume	Rotten eggs

Please explain why you chose this sample.

_____

_____

_____

_____

_____

_____

_____

Explain how odors spread through a room.

_____

_____

_____

_____

_____

_____

_____

The Nose Knows

The following articles can also be found by clicking the computer in the CD-ROM laboratory for Exploration 1:

- *Diffusing the Confusion*
- *Making Sense of Scents*

Diffusing the Confusion

What Is Diffusion?

All matter is made up of particles that are in constant motion. Even the air around us consists of billions of particles that are moving at high speeds in random directions. This characteristic of matter allows diffusion to take place. **Diffusion** is the process in which particles of one substance move from an area of higher concentration to areas of lower concentration.

The particles of food coloring are moving from an area of higher concentration to areas of lower concentration.

How does this work? Look at the photo showing what happens when a drop of food coloring is added to water. The food coloring is an area of high concentration (of food coloring particles),

and the water is an area of low concentration (of food coloring particles). As soon as the drop of food coloring hits the water, the particles of food coloring immediately move toward areas of lower concentration in the water. After a period of time, the particles reach a relatively uniform concentration, and diffusion ceases.

Snowball Diffusion

To make diffusion easier to understand, imagine that Alan and Byron are having a snowball fight across the fence that separates their backyards. Each boy is allowed to throw a set number of snowballs, say one-tenth of them, into the other boy's yard every minute.

Suppose that at the beginning of the fight Alan has 100 snowballs in his yard and Byron has 50 snowballs in his yard. Alan's yard is the area of higher concentration, and Byron's yard is the area of lower concentration. In the first minute, Alan throws 10 snowballs (one-tenth of 100) into Byron's yard and Byron throws 5 snowballs (one-tenth of 50) into Alan's yard. At the end of the first minute, Alan has 95 snowballs in his yard and Byron has 55 snowballs in his yard. Alan has a decreased concentration of snowballs, and Byron has an increased concentration of snowballs. Therefore, there is a net movement of snowballs from the area of higher concentration to the area of lower concentration.

During the second minute, Alan again throws one-tenth of his snowballs into Byron's yard and Byron again throws one-tenth of his snowballs into Alan's yard. At the end of the second minute, Alan has 91 snowballs in his yard and Byron has 59 snowballs in his yard. As time passes, Alan loses snowballs while Byron gains snowballs because the net movement of snowballs continues to be from Alan's yard to Byron's yard.

As Alan and Byron continue to play, they move toward having an average of 75 snowballs each. At this point, there is a uniform concentration of snowballs in each yard. If they continued playing, on average they would each throw the same number of snowballs every minute. There would be no net movement of the snowballs, and the process of "snowball diffusion" essentially stops.

Different Rates of Diffusion

In general, diffusion occurs more readily between gases and between liquids than it does between solids. This is because the particles that make up gases and liquids are farther apart and move faster than the particles in solids. Although it may be difficult to imagine, diffusion does take place between solids. For example, if samples of zinc and copper are clamped together for several months, a small amount of diffusion between particles of zinc and copper will occur.

Concentration

Differences in concentration affect the rate of diffusion. In general, diffusion occurs more rapidly when there is a large difference in concentration between two areas. Refer back to the example of the food coloring in the beaker of water. The difference in concentration (of food coloring particles) between the drop of food coloring and the water is greatest when the drop of food coloring first enters the water. Initially, diffusion occurs very rapidly. As diffusion continues, the difference in concentration between the water and the food coloring lessens, and diffusion begins to slow down. Once the food coloring particles are uniformly distributed throughout the water and there is little difference in concentration, then diffusion effectively stops.

Temperature

Differences in temperature also affect the rate of diffusion. When the temperature of a substance is raised, the molecules move faster and rebound farther after collisions due to an increased amount of kinetic energy. Consider how much faster a spoonful of sugar dissolves in a cup of hot tea than it does in a glass of iced tea.

Pressure

The rate of diffusion is also affected by pressure. Under high pressure, particles are squeezed closer together. As a result, there is less time between particle collisions, and particles get sent in new directions at a faster rate. This increased rate of collisions spreads the particles out faster, increasing the rate of diffusion.

Osmosis—Diffusion Through a Membrane

In living systems, water particles often diffuse through a semipermeable membrane, such as a cell membrane. A **semipermeable membrane** allows certain particles to pass through it while blocking others. In general, semipermeable membranes prevent the passage of larger particles. The process of diffusion through a semipermeable membrane is called **osmosis.**

The concentration of water inside a cell affects whether water moves into or out of the cell through the cell membrane. When there are more water particles outside the membrane than there are inside the membrane, water moves into the cell. When there are fewer water particles outside the membrane than there are inside the membrane, water moves out of the cell. When the concentration of water particles is equal on either side of the membrane, there is no net movement of water particles in either direction.

Making Sense of Scents

Wow, Do You Smell!

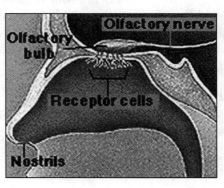

Your olfactory system allows you to recognize a variety of odors—from the aroma of fresh-baked cookies to the stink of rotten garbage. You sense these smells by inhaling particles that have diffused from their source into the air. Receptor cells inside your nose react to these particles by sending a message along the olfactory nerve to the olfactory bulbs in the brain. There, the messages are interpreted into the sensation of smell.

Exploration 1

Smelling is not the only purpose your olfactory system serves, however. It also plays an important function in your sense of taste. If you have ever held your nose to eat or drink something you dislike, cold medicine, for instance, then you have experienced first-hand the effects of your olfactory system on your sense of taste. In fact, if you were blindfolded and wearing noseplugs, you probably could not tell an apple from an onion just by tasting them.

It Is a Diffusing World

As you know, many things have a scent, and these scents can provide valuable information. Many animals rely heavily on their sense of smell to find food, shelter, and mates. Some animal smells can also serve as warnings. For example, skunks are famous for their smelly emissions. A skunk will "spray" if it feels threatened. This offensive tactic helps the skunk thwart other animals.

Smell also plays an important role in human behavior. From infancy, you develop definite opinions about different odors. As your brain learns to recognize different odors, it can determine safe smells from dangerous smells. For example, you might not drink sour milk because particles diffusing from the milk let you know that the milk is not safe to drink. In this way, your sense of smell helps you make sensible decisions.

Smelly Chemicals

Most people can recognize thousands of different objects just by their smells. The distinctive scent of a substance is caused by different chemicals or combinations of chemicals.

Food

Certain chemicals give foods their distinctive smells. For example, butyl acetate makes an apple smell like an apple, while ocytl acetate makes an orange smell like an orange. The names of some scent-producing chemicals sound like the substance that produces the odor. For example, cinnamaldehyde is the chemical that gives cinnamon its fiery smell, and vanillin gives the vanilla bean its odor. Other names of smelly chemicals are derived from the scientific name of the substance. For example, the scientific name for garlic is *Allium sativum,* and the chemical that gives garlic its smell is called allicin.

Alcohol

If you open a bottle of rubbing alcohol, you will probably be able to smell the alcohol almost immediately. This is because rubbing alcohol diffuses very quickly. Rubbing alcohol is often used as a solvent because it reacts with many different kinds of chemicals. For example, one way to remove ink from skin and clothing is by applying rubbing alcohol to the stain. Alcohol is also highly flammable. If a flame gets anywhere near the quickly diffusing particles of alcohol, you will witness a sure-fire reaction.

Perfume

Perfumes may contain over 100 different ingredients. The most familiar ingredients come from fragrant plants or flowers, such as sandalwood or roses. Other perfume ingredients come from animals and from human-made chemicals. Some substances, like civet musk, are used to make the odors in the perfume last longer. Some perfumes sold in stores are synthetic, which means that the chemicals were created and mixed in a laboratory. For example, a combination of geraniol and beta-phenyl ethyl alcohol produces a scent that smells like roses. Many perfumes, especially those packaged in spray bottles, contain isopropyl alcohol, which allows the scent to diffuse quickly. However, this ingredient also makes the perfume flammable.

Hydrogen Sulfide

Some dangerous chemicals produce distinct odors. Hydrogen sulfide, for example, is a toxic, gaseous chemical that smells like rotten eggs. The brain recognizes the smell of hydrogen sulfide as unpleasant, and we instinctively want to get away from the smelly source. However, not all dangerous substances produce a smelly warning. Natural gas, for example, has no odor, so gas companies mix odor-causing chemicals with the gas. This way, you can detect natural gas leaks before they reach toxic or explosive levels in your home.

Sea the Light

Key Concepts	Materials can differ in density because of their structure at the particle level. An underwater lamp that is neutrally buoyant will neither sink nor rise in sea water.
Summary	Diane Sittie, a scuba-diving enthusiast, wants to create an underwater lamp that she can use where she dives. She has sent a prototype lamp base and some ballast disks to Dr. Labcoat's lab. Ms. Sittie needs to know which ballast disk to add to the lamp base so that the entire lamp will be neutrally buoyant in the sea water where Ms. Sittie dives.
Mission	Recommend a metal ballast disk for an underwater hanging lamp.
Solution	A titanium ballast disk used in conjunction with the waterproof lamp base will result in an underwater hanging lamp that has a total density closest to that of the water where Ms. Sittie dives. As a result, the lamp will be neutrally buoyant.
Background	Scuba diving requires special equipment. The most obvious part of the equipment is the breathing apparatus (*scuba* stands for *s*elf-*c*ontained *u*nderwater *b*reathing *a*pparatus). To allow a diver to breathe underwater, a tank of compressed air, usually consisting of a mixture of helium, oxygen, and nitrogen, is attached to hoses and regulators. One regulator is connected to the tank. Called a *first-stage regulator,* it controls how much compressed air flows from the tank through the hoses, and how compressed the air is. This regulator allows a diver to decompress as he or she surfaces, ensuring a safe transition from breathing compressed air to breathing normal air. Another regulator, called a *second-stage regulator,* is attached to the diver's mouthpiece. This regulator controls the opening and closing of a mechanism in the mouthpiece that determines how much force the diver must use to inhale and exhale.

Another important piece of equipment used by scuba divers is a buoyancy control device, or BCD. BCDs are usually similar to vests or backpacks; they are designed to hold the air tank and other accessories. BCDs also allow divers to control their buoyancy underwater. By inflating the BCD, the diver rises toward the surface, and by deflating the BCD, the diver sinks. The inflation and deflation can be done orally or by an automatic inflator. Because the human body tends to float rather than sink, scuba divers in ocean water also use weights as ballast. The weights allow them to descend to certain depths as well as give them stability and mobility underwater. Some BCDs have pockets designed to hold these weights, and some divers wear weight belts. |

Exploration 2

Teaching Strategies

Because students may have difficulty connecting the concept of density with the particle theory of matter, you may need to discuss how the atomic models on the back counter of Dr. Labcoat's lab demonstrate density on the particle level. Make sure students understand that these atomic models show how much matter occupies a given space (density) for a given element. You may want to use the periodic table to explain that elements with higher atomic numbers can sometimes also have greater densities because the atomic number indicates the number of particles (protons) within an atom of a given element. You can also use the atomic models to explain how particle arrangement (how close together or far apart particles are within a given molecule) can also determine the density of a substance.

To help students understand how the density of fluids can affect buoyancy, you may want to conduct a demonstration that shows how fluids with greater densities can exert greater buoyant forces. Mix a solution of salt and water that is concentrated but not saturated and that remains clear in color. Pour this solution into a glass or a beaker. Fill an identical glass or beaker with fresh water. Place a whole raw egg in each water sample. The egg placed in the salt water should float, and the egg placed in the fresh water should sink. Explain this phenomenon to students in terms of buoyant force and density.

Bibliography for Teachers

de Grasse Tyson, Neil. "On Being Dense." *Natural History,* 105 (1): January 1996, pp. 66–67.

Peterson, I. "Explosive Expansion of Atomic Nuclei." *Science News,* 147 (15): April 15, 1995, p. 228.

Bibliography for Students

Hoover, Pierce. "The Deep." *Popular Mechanics,* 173 (1): January 1996, p. 68.

Surkiewicz, Joe. "Derby of the Deep." *Boys' Life,* 86 (7): July 1996, p. 42.

Other Media

The Atom
Video and book
SVE (Society for Visual Education)
6677 N. Northwest Highway
Chicago, IL 60631
800-624-1678

In addition to the above video, students may find relevant information about the particle theory of matter, density, and buoyancy by exploring the Internet. Interested students can search for articles under *chemistry* using keywords such as *atoms; density; protons, neutrons, and electrons;* and *atomic models.* Students could also search under *physics* to find information about *buoyancy.* Students can also access information about *scuba diving, scuba gear,* and *scuba certification* on the Internet.

Sea the Light

1. Ms. Sittie wants to create an underwater hanging lamp. What help does she need from you?

   _____

   _____

   _____

   _____

2. What does *ballast* mean? (If you aren't sure, use the CD-ROM articles to help you.)

   _____

   _____

   _____

3. What purpose do you think the ballast disks serve in the design of the underwater lamp?

   _____

   _____

   _____

   _____

4. Describe how you will use the equipment on the lab's front table to answer Ms. Sittie's questions.

   _____

   _____

   _____

   _____

   _____

   _____

Exploration 2

5. Use the equipment to conduct all of the necessary tests, and record your data in the first two columns of the table below. Then use your results to calculate the values for the third column.

Ballast disk	Mass (g)	Volume (mL)	Density (g/mL)
A Copper			
B Aluminum			
C Brass			
D Titanium			
E Platinum			
F Zinc			

6. Calculate the total density of the entire lamp for each individual ballast disk. (If you aren't sure how to calculate the density of an object with multiple parts, examine the CD-ROM articles.)

Ballast disk	Density
A Copper	
B Aluminum	
C Brass	
D Titanium	
E Platinum	
F Zinc	

7. Examine the materials on the back counter of the lab. Use what you see to explain why the different ballast disks have different densities.

_____

_____

_____

_____

_____

_____

_____

_____

8. What is buoyant force? (If you're not sure, check out the CD-ROM articles.)

_____

_____

_____

_____

_____

_____

_____

_____

9. Describe the differences among underwater objects that are positively, negatively, and neutrally buoyant. (Hint: Check out the CD-ROM articles.)

_____

_____

_____

_____

_____

_____

_____

_____

_____

_____

Record your answers in the fax to Ms. Sittie.

Exploration 2

Name ________________________ Date ____________ Class ____________

Exploration 2
Fax Form

FAX

To: Ms. Diane Sittie (FAX 817-555-4459)

From:

Date:

Subject: Sea the Light

Please complete the following chart:

METAL	MASS	VOLUME	DENSITY
Aluminum			
Brass			
Copper			
Platinum			
Titanium			
Zinc			

What is the density of the waterproof lamp base?

____

Please indicate your metal selection for the ballast disk here: ____________________

Why did you pick this metal?

____

____

____

____

____

____

Sea the Light

The following articles can also be found by clicking the computer in the CD-ROM laboratory for Exploration 2:

• *A Dense Discussion*
• *Oh, Buoy!*
• *Diver Down*

A Dense Discussion

What Is Density?

Density is a measure of the amount of matter in a given amount of space. Density is also defined as mass per unit volume. Within a given volume, dense materials have a lot of mass, whereas less-dense materials have less mass. For example, 5 mL of steel, a fairly dense material, has more mass than does 5 mL of cork, a less-dense material.

Because density is a physical property of a substance, objects can be distinguished by their densities. One way to distinguish between liquids, for example, is by a technique called layering. A liquid that is denser than another will sink below the less-dense liquid. For example, corn oil floats on top of water because corn oil is less dense than water.

Calculating Density

To determine the density of an object, you must first measure its mass and its volume. A balance can be used to measure the object's mass. A graduated cylinder can be used to measure the object's volume. The object's density can then be calculated using this simple equation:
density = mass ÷ volume.

Suppose that you have a sample of lead that has a mass of 34 g and a volume of 3 mL. Inserting these measured values into the equation, you get the following:
density = 34 g ÷ 3 mL = 11.35 g/mL.

How can you determine the overall density of an object that consists of two parts with different densities? One way to do this would be to

measure the mass and volume of each part and then use the following equation: total density = (mass of first part + mass of second part) ÷ (volume of first part + volume of second part).

Density Table

Here are density values for some common materials, including solids, liquids, and gases.

Material	Density (g/mL)
Helium (gas)	17.85×10^{-5}
Oxygen (gas)	13.31×10^{-4}
Ethyl alcohol (liquid)	0.79
Water (liquid)	1.00
Iron (solid)	7.87
Silver (solid)	10.50
Lead (solid)	11.34
Mercury (liquid)	13.55
Gold (solid)	19.31

Particles and Density

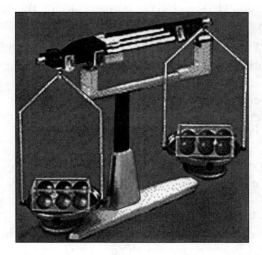

Different substances have different densities. This is partly due to the different masses of the particles that make up the substances. Imagine that you have several solid lead balls and several solid rubber balls, all of the same size. Now suppose that you fill one box with the lead balls and another box with the rubber balls. The balls in each box are packed identically and are the same size. As you can probably imagine, the box full of lead balls is heavier. Therefore, you can conclude that the box filled with lead balls is more dense than the box filled with rubber balls.

Density also depends on the structure of a material. A substance that has a great deal of space between its particles will be less dense than a substance with tightly packed particles. For example, graphite and diamond are both made of carbon. The carbon particles in each substance have the same chemical makeup and are the same size. However, the carbon particles in diamond are packed together much more tightly than they are in graphite. As a result, diamond is more dense than graphite.

Conditions That Influence Density

Certain conditions can influence the density of a substance. For example, a change in temperature can change the density of a substance. As a substance's temperature increases, the substance usually expands, or takes up more space, filling a larger volume. As a result, the density of the substance decreases because the same mass occupies a larger volume. On the other hand, a substance usually contracts as its temperature decreases. As a result, its density increases because the same mass occupies a smaller volume. Water is an exception to the effect of temperature on density. When water freezes, it expands rather than contracts. But we will talk about this later.

Another condition that affects the density of a substance is pressure. For example, if you squeeze a sponge, the particles of the sponge squeeze closer together, taking up less volume. This makes the density of the sponge increase. If you let go of the sponge, the pressure is removed and the particles return to their original placement, increasing volume and decreasing density.

A Sea of Different Densities

The composition of sea water varies around the world. Some regions of the ocean have a relatively high salt content; other regions have a relatively low salt content. At a given temperature and pressure, sea water with a high salt content is more dense than sea water with a low salt content.

Temperature and pressure also affect the density of sea water. Cold water, such as that found near the ocean floor and in the polar regions, is generally more dense than warm surface water or water found in tropical areas. The density of sea water also increases with depth because pressure increases with depth.

Oh, Buoy!

A Force in Fluids

Why do some things float while others sink? Well, fluids exert an upward force, called a **buoyant force,** on objects that are partially or completely submerged in them. If you have ever floated on an air mattress in a swimming pool, you've experienced this buoyant force. You can also feel it when you lift a heavy object, such as a brick, underwater. The brick seems a lot lighter underwater than it would above water. That's because the water exerts an upward buoyant force on the brick that makes the brick seem like it weighs less.

Buoyant Force (Part 1)

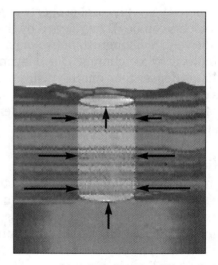

Buoyant force is due to the difference in the amount of fluid pressure between the top and the bottom of this column of water.

Perhaps you are wondering where buoyant force comes from. Look at the picture of a column of water in a lake. The bottom of the column receives greater pressure than does the top of the column because the bottom is supporting a larger amount of water above it.

The difference in the amount of force due to pressure on an object is known as the **buoyant force.** You may wonder why the buoyant force is always in the "up" direction. The buoyant force is the result of a difference in pressure. If an object is squeezed equally in every direction, it will not feel a force pushing it up, down, or sideways. Because the pressure in a fluid increases with depth, the bottom of an object in a fluid always receives more pressure than the top of the object. This is not the case for the sides of an object. When there is pressure on one side of an object, then there is equal pressure (at the same depth) on the other side of the object. Therefore, the only difference in pressure is from top to bottom. Buoyant force is an upward force because the pressure at the greater depth is always greater than the pressure at the lesser depth.

Buoyant Force (Part 2)

Now let's suppose that the underwater column is a tennis-ball canister. Again, the pressure at the bottom of the canister is greater than the pressure at the top of the canister, so there is an upward buoyant force. But now we have to consider something else: the tennis-ball canister displaces an amount of water equal to its own volume. Archimedes, a Greek scientist and mathematician, discovered that the buoyant force on an object is equal to the weight of the fluid displaced by the object. In this example, the buoyant force equals the weight of the water displaced by the tennis-ball canister.

The weight of the object itself does not determine the buoyant force. An object's weight comes into play only when the effect of the buoyant force on the object is considered. In other words, we consider the object's weight when we're trying to figure out whether an object will sink or float.

Buoyancy and Archimedes

An object immersed in a fluid experiences a buoyant force that is equal to the weight of the fluid displaced by the object. By this reasoning, an object will float if, and only if, it can displace a volume of liquid that is equal to its own weight.

Let's consider a steel ball and a steel boat, each with the same mass. Both are made of the same amount and type of material, but as you can probably guess, the boat will float and the ball will

sink. Why? The difference is in the amount of water displaced by each object. The boat can displace a volume of water that weighs the same as the boat weighs. But because of its shape, the ball does not displace much water. Because the ball does not displace a volume of water that is equal to its own weight, it sinks.

Buoyancy and Density

An olive floats in the liquid in beaker A but sinks in the liquid in beaker B.

What would happen if you put identical objects in different liquids? Take a look at the olives in the beakers shown here.

In this example, each olive has the same volume and the same mass. So why does one olive float while the other one sinks? The answer has to do with the liquids in which they are placed. One liquid is more dense than the other. Can you figure out which one?

Remember that the buoyant force on an object is equal to the weight of the fluid it displaces. This being the case, the liquid in beaker A must be more dense than the liquid in beaker B. Why? If each olive displaces the same volume of liquid, the liquid in beaker A must weigh more. The only way it can weigh more is if it is more dense.

The Density of Water

Usually, solids are more dense than liquids because the particles in solids are closer together. So why does ice float in water?

The way in which individual water molecules are arranged affects the water's density. In a liquid state, the molecules are relatively close together. In a solid or frozen state, the molecules are locked into positions that are farther apart than the molecules are in the liquid state. As a result, the volume of the water increases when it freezes while the mass stays the same. That's why ice floats in water.

Controlling Buoyancy

How does a submarine control whether it sinks, floats, or stays in the same position underwater? The trick is to control the density of the submarine.

Built into the submarine are tanks that can be filled with water to give the submarine ballast. **Ballast** is a term used to describe anything heavy that is carried in ships or other structures to add mass. To make the submarine submerge, the tanks are filled with sea water. This addition of mass (without a change in the volume of the submarine) makes the submarine more dense. As a result, it sinks and can be described as **negatively buoyant.** When the submarine sinks to the depth at which its density equals the density of the surrounding water, it becomes **neutrally buoyant.** An object is neutrally buoyant when it neither sinks nor rises underwater. To bring the submarine to the water's surface, the water is pumped out of the tanks (and the tanks are filled with air), decreasing the overall density of the submarine. When its density is less than the density of the surrounding sea water, the submarine rises and can be described as **positively buoyant.**

Diver Down

A Brief History of Underwater Diving

For over 2000 years, people have attempted to design mechanical means to stay underwater for as long as possible. Take a look at the paragraphs below for some momentous events in the history of underwater diving.

332 B.C.—Aristotle, the Ancient Greek philosopher, describes a diving bell used by his prize student, Alexander the Great.

1500s—Renaissance artist and inventor Leonardo da Vinci designs a single diving system that combines air supply and buoyancy control.

1808—Friedrich von Drieberg invents the Triton apparatus, which features a backpack air reservoir supplied with air from above water. By nodding back and forth, the diver receives air through a valve.

1825—William James designs a diving system of closed tanks with compressed air. (The word *scuba* is an acronym for self contained underwater breathing apparatus.)

1911—The Davis False Lung, invented by Sir Robert Davis, saves lives worldwide in emergency submarine rescues. The device is a self-contained "rebreather" that provides crew members with enough air to swim to the water's surface.

1937—At the Paris International Exposition, divers demonstrate a scuba system that combines compressed air tanks with a valve that lets them regulate the amount of air they take in.

1943—Jacques Cousteau and Emile Gagnan invent the aqualung, the first safe and simple underwater breathing device.

Spotlight on Scuba Diving

In the 1940s and 1950s, Jacques Cousteau and his friends helped popularize scuba diving all over the world. His underwater films have provided a glimpse of the world that exists below the surface of the water.

In the United States, scuba-diving schools can teach you to use diving equipment safely. Minimum age requirements vary by scuba school and by course, but most require that you be at least 12 years old to become a junior certified diver. Open Water Diver is the most common certification for beginning scuba divers. It usually requires a minimum of 12 classroom hours, 12 hours of confined water training (in a swimming pool), and four to five open-water or ocean dives. Open Water Diver certification is a prerequisite for more-advanced diving courses, including Advanced Scuba Diver and Divemaster courses. Advanced scuba students can learn underwater photography, rescue diving, deep diving, cave diving, and search-and-recovery diving.

Stranger Than Friction

Key Concepts	Frictional force is the force that opposes motion when two objects are touching. The force of friction is not dependent on surface area.
Summary	Mr. Norm N. Cline has designed a new ride for his amusement park. The ride consists of a slide and several toboggans. He wants to know what material he should use to construct the slide and the bottom of the toboggans so that the amount of friction between the two will ensure a ride that is both exciting and safe. He also wants to know what size to make the toboggans.
Mission	Help the owner of an amusement park choose the best materials for a park ride.
Solution	Constructing the slide and the bottom of the toboggans out of stainless steel results in an amusement park ride that is both exciting and safe. The size of the toboggan has no effect on the slide's performance because frictional force is determined only by the normal force and the coefficient of friction between two surfaces, not by surface area.
Background	Many roller coasters have only one motorized mechanism—the conveyor belt that brings the train of cars to the top of the first hill. From there, only the force of gravity and the frictional force between the track and the wheels of the cars determine how fast the cars go down the hills, through the loops, and along the track. The taller a roller coaster track is at its initial height, the faster the velocity that the cars can achieve as they are accelerated by the force of gravity on their way down the first hill. Engineers continue to improve on the design of roller coasters to make them even more thrilling, more extreme, and faster. Some rides can send passengers zooming around at over 120 km/hr!
	There is no such thing as a 100-percent-efficient machine, and that certainly influences how many roller coasters are designed. At the crest of the first hill, which is the tallest point on many roller coasters, the potential energy of the coaster cars is greater than it will be at any other point during the entire ride. As the cars head down the first hill, the potential energy is converted into kinetic, heat, and sound energy. The total amount of energy is conserved, but only the kinetic energy can be converted back into potential energy. This kinetic energy is enough to carry the cars up another hill. However, because the amount of the cars' kinetic energy is less than the original amount of potential energy, the second hill cannot be as high as the first hill.

Exploration 3

Teaching Strategies

The main goal of this Exploration is to help students understand what causes frictional force and how frictional force affects moving objects. To ensure that students understand how the coefficient of friction functions in determining frictional force, discuss as a class the information in the CD-ROM articles related to the mathematical expression for the coefficient of friction. Determining frictional force mathematically, the coefficient of friction (μ) is the multiplier value for the normal force (F_n) because $F_f = \mu F_n$. This may help students understand how smaller coefficients of friction result in smaller frictional forces for a given normal force. For example, a coefficient of friction of 0.70 and a normal force of 10 N produce a frictional force of 7 N ($F_f = 0.70 \times 10$ N $= 7$ N). If the coefficient of friction is 0.30 and the normal force remains 10 N, then the frictional force is 3 N ($F_f = 0.30 \times 10$ N $= 3$ N).

Another concept students must understand is that the surface area of materials in contact with one another has no bearing on frictional force. By exploring the equipment on the back counter in the lab, students should be able to grasp this concept and then apply it to what the different-sized toboggans demonstrate. Once students understand that surface area does not affect frictional force, they should realize that conducting experiments with every possible combination of materials and toboggan size is not necessary.

Bibliography for Teachers

"Coast Coaster." *U.S. News & World Report,* 120 (23): June 10, 1996, p. 21.

Wade, Bob. "Hot Wheels in the Laboratory." *The Physics Teacher,* 34 (3): March 1996, p. 150.

Young, Janet. "Complex Creations from Simple Machines." *Science Teacher,* 61 (1): January 1994, pp. 16–19.

Bibliography for Students

Koehl, Carla, and Sarah Van Boven. "Better and Faster Ways to Lose Your Lunch." *Newsweek,* 127 (22): May 27, 1996, p. 8.

Timney, Mark C. "Ups and Downs of Coaster Physics." *Boys' Life,* 86 (6): June 1996, p. 50.

Other Media

Energy at Work
Video or videodisc
Churchill Media
6677 N. Northwest Highway
Chicago, IL 60631
800-829-1900

In addition to the above video and videodisc, students may find relevant information about work and energy by exploring the Internet. Interested students can search for physics articles using keywords such as *work, energy,* and *machine efficiency.* Students can also access information about *roller coasters* by exploring the Internet.

Stranger Than Friction

1. Mr. Cline is hard at work on his design for a new amusement park ride. What information is he seeking from you?

_____

_____

_____

_____

2. Describe the equipment Dr. Labcoat has set up on the front lab table and the back counter.

_____

_____

_____

_____

3. What is frictional force, and how does it affect a moving object? (Hint: If you're not sure, check out the CD-ROM articles.)

_____

_____

_____

_____

4. What does the wall chart in the lab show about normal force and the coefficient of friction?

_____

_____

_____

_____

_____

_____

Exploration 3

5. Use the force meter on the back lab counter to find the force required to pull each block. Record your results below.

_____

_____

_____

6. Does the amount of surface area touching the block affect the force required to pull it? Why or why not?

_____

_____

_____

_____

_____

7. Conduct the necessary tests with the prototype for the Camelback Super Slide, and record your results in the table below. (Hint: It may not be necessary to try every possible combination.)

Material component for slide	Material component for toboggan	Toboggan size (cm)	Observations

Record your answers in the fax to Mr. Cline.

FAX

To: Mr. Norm N. Cline (FAX 281-555-5276)

From:

Date:

Subject: Stranger Than Friction

What material do you recommend for the construction of the slide?

_____

What material do you recommend for the construction of the toboggan?

_____

What is your recommendation regarding toboggan size?

100 cm
120 cm
140 cm
any of the above

What effect does the size of the toboggan have on the performance of the Camelback Super Slide? Explain.

_____

_____

_____

_____

_____

_____

The Nature of Friction

Friction Fundamentals

Imagine you are pushing a box loaded with books across the floor. As you push, the floor seems to oppose the motion of the box. This opposition of motion is due to the friction between the box and the floor. **Frictional force** is the force that opposes motion between two surfaces that are touching. In order to slide the box of books across the floor, you have to overcome friction. When the force of your push on the moving box matches the force of friction, the box moves at a constant velocity; that is, the box is not accelerating due to unbalanced forces. The box continues to move at this velocity until you change the amount of force applied. If you stop pushing the box, the box stops because frictional force is no longer opposed.

The amount of friction between two surfaces depends on two factors: the type of surfaces that are in contact and how hard the surfaces are pushed together. The more books you load into the box, the more difficult it is to push the box across the floor. Likewise, if the floor is covered with rough carpet instead of with smooth wood, the box is harder to push. Frictional force does not usually depend on the size of the surfaces in contact or on how fast the surfaces are moving.

Different Kinds of Friction

Have you ever noticed that it is harder to start an object in motion than it is to keep it in motion? Think about sliding down a slide. At first you seem to "stick" to the slide, so you have to apply a large force to get started. This is because initially you must overcome the frictional force caused by static friction. **Static friction,** or starting friction, is the friction between objects that are in contact but not yet in motion. Once you overcome static friction, you are able to slide down easily. This is because once you start moving, the larger static frictional force gives way to the lesser kinetic frictional force. **Kinetic friction** is the friction between moving objects that are in contact. The kinetic frictional force is almost always less than the static frictional force.

The amount of kinetic friction between objects varies, depending on how the objects are touching. For example, sliding a box across a floor is an example of overcoming sliding kinetic frictional force. Putting wheels under the box would make the box easier to push. This is because you have changed the nature of the friction from sliding kinetic friction to rolling kinetic friction. Of course, the friction between the floor and the box isn't the only kind of friction that comes into play. Friction also exists between the box and the air. This type of friction is called fluid kinetic friction.

Decreasing and Increasing Friction

Sometimes it is desirable to decrease the amount of friction between two surfaces. For example, adding oil, grease, or other lubricants to a car engine is a way to reduce friction between the engine's moving parts. Reducing friction in a car engine protects the engine and keeps it from getting too hot. Another example of reducing friction is waxing downhill skis. The wax allows the skis to slide along the snow with less resistance, increasing the skier's maximum acceleration down a slope.

Other times, it is desirable to increase friction between surfaces. This is particularly useful when you want to stop an object like a car or a bicycle. The brakes on both of these vehicles create friction between brake pads and the turning wheels. When you pull on the brake lever of a bicycle, you are pressing rubber brake pads against the rim of the wheel. As a result, the wheel slows down.

Friction and Surface Area

How fast the brake pads stop the bicycle described in the previous section certainly depends on how hard you squeeze the brake lever. The stopping force also depends on the material from which the brake pads are made. Fresh new brake pads would provide a greater frictional force than old, hard brake pads. So how about larger brake pads? Will larger brake pads provide more frictional force than smaller ones? Believe it or not, the answer is no. Friction is not affected by the area of contact between surfaces.

Machines and Friction

Frictional force affects the operation of every machine. In fact, friction is necessary for many machines and mechanical systems to work at all. For example, the friction between a nail (a wedge) and the surface into which it is hammered, such as wood, keeps the nail in a sturdy position. Frictional force can also work against a machine. For example, kinetic frictional force opposes the motion of a box moving up an inclined plane.

Frictional forces affect the efficiency of machines. Some of the energy put into a machine or mechanical system is converted into heat and sound energy by the friction of moving parts. That means that large frictional forces can decrease the amount of energy available to do work, whereas smaller frictional forces can result in more energy being available to do work. In other words, by reducing friction, a machine's efficiency can be increased.

Friction by the Numbers

Important Forces

For a closer look at frictional force, consider a picture representation of what goes on when a box moves across a surface.

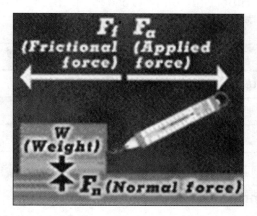

One obvious force involved in this situation is the weight of the box. Weight is a downward force, but the box is not accelerating downward, so there must be another force that opposes weight. This force is called the **normal force (F_n).** It is the perpendicular force pressing the two objects together. The greater the normal force on an object, the greater the frictional force. Because this box is on a horizontal surface, the normal force is equal in magnitude to the weight of the box. Notice that the arrows representing the applied force and the frictional force point in opposite directions. That's because frictional force opposes the motion of an object.

Coefficients of Friction

Remember that the types of surfaces in contact are very important in determining the amount of frictional force on an object. The friction between two surfaces can be expressed as a ratio. This ratio is called the coefficient of friction. The **coefficient of friction (μ)** is the ratio between the frictional force and the normal force. The amount of friction between two objects depends on whether an object is moving or starting in motion. Remember that it is more difficult to start an object in motion than it is to keep a moving object moving. We can express the difference in the amount of frictional force by using a coefficient of static friction and a coefficient of kinetic friction. The chart on the next page lists values for the coefficients of friction for various surfaces.

Exploration 3

Coefficients of Friction

Surfaces	Coefficient of static friction*	Coefficient of kinetic friction*
glass on glass	0.94	0.40
rubber tire on dry road	0.90	0.70
rubber tire on wet road	0.70	0.50
steel on ice	0.02	0.01
steel on steel	0.74	0.57
steel on steel with lubricant	0.20	0.12
teflon on teflon	0.40	0.04
wood on wood	0.50	0.40

*The values shown are averages.

Comparing the Numbers

A high coefficient of friction indicates a greater resistance to movement. A low coefficient of friction indicates a lower resistance to movement. In general, for a given pair of surfaces, the static coefficient of friction is greater than the kinetic coefficient of friction.

An Equation and Its Parts

Once you know the normal force and the coefficient of friction for a pair of objects, you can determine the frictional force. The equation for frictional force looks like this:

frictional force (F_f) = coefficient of friction (μ) $\times$ normal force (F_n)

Suppose that a wooden block is sitting on a horizontal wooden board and that you want to know how much force is necessary to get the block going. Obviously, you need to figure out what the starting frictional force is because that is the force that you will have to overcome.

To find the static frictional force, you would look up the coefficient of static friction for wood on wood, which is 0.50. Now you need to know the normal force. Because the block is horizontal, the normal force is equivalent to the block's weight. Suppose that value is 100 N. Now you have the values you need to solve the equation and find the force of static friction.

Given:
$\mu = 0.50$
$F_n = 100$ N

Inserting these values into the equation $F_f = \mu \times F_n$, you have:

$$F_f = 0.50 \times 100 \text{ N} = 50 \text{ N}$$

Thus the force of static friction for the given wooden block on the given wooden surface is 50 N. That means you have to apply a force greater than 50 N in order to set the block in motion.

That's Amusing!

Record-Holding Roller Coasters

Engineers are constantly creating improved designs for rides that are guaranteed to take your breath away. Going through a loop, leaning through a tight turn, and plummeting from death-defying heights at top speeds all add to the excitement of riding roller coasters.

You can rank roller coasters any number of ways—from the fastest to the steepest to the tallest. But which coasters are the scariest? the most thrilling? Well, that is entirely up to you!

Steepest

The Ultra Twister, a steel roller coaster at Astroworld, in Houston, Texas, is the steepest roller coaster in the world. It drops you at an angle of almost 90 degrees from horizontal. This ride is guaranteed to make you want to hold on to your hat!

Highest

At almost 80 m, the Megacoaster at Fujiyama Park, in Japan, is the highest roller coaster in the world. It also has the longest drop of any roller coaster.

Fastest

The Steel Phantom at Kennywood, in Pennsylvania, and the Desperado at Buffalo Bill's Resort, in Nevada, are tied for the title of fastest roller coaster in the world. (They are also tied for highest roller coaster and longest drop in the United States.) These coasters travel at top speeds of about 128 km/hr. That's faster than the legal speed limit on most highways!

Exploration 3

Latitude Attitude

Key Concepts	The Coriolis effect explains the apparent deflection of moving objects relative to the surface of the Earth. Airplane pilots must adjust their flight paths to account for the Coriolis effect.
Summary	Capt. Corey O. Lease has been commissioned to deliver supplies to Antarctica, where a team of scientists are studying the Mertz Glacier. Capt. Lease must make the trip immediately, but because the weather conditions there are terrible, she will not be able to land her plane at the site. Instead, she must drop the supplies as she flies over the area. Capt. Lease needs to know the best flight direction and average speed to use so that the supplies arrive safely at the site.
Mission	Advise Capt. Lease of the best flight plan to successfully complete her mission.
Solution	Flying at a compass direction of 230° and a speed of 725 km/hr is the best plan for Capt. Lease. Flying at a compass direction of 220° and a speed of 850 km/hr also allows Capt. Lease to deliver the supplies, but this combination uses too much fuel.
Background	The spherical shape of the Earth and the properties of centripetal acceleration contribute to the Coriolis effect. Centripetal acceleration is the acceleration needed to keep an object moving in a circle at a given radius. For a given centripetal acceleration, if the velocity of the spinning object increases, the radius of rotation increases as well ($a = v^2/r$). If the velocity of the object decreases, the radius decreases.
	The centripetal acceleration of objects on the Earth is caused by gravitational force. The radius of centripetal acceleration for the Earth is measured as a perpendicular line from the Earth's axis of rotation to any given point on the Earth's surface. As a result, there are different radii of centripetal acceleration for different points on the Earth's surface. For example, a point on the equator has a longer radius of centripetal acceleration than does a point near one of the poles. At a given radius, a point has a given velocity. For example, a point at 30°N latitude is rotating east at a velocity of 1400 km/hr. Suppose an object at this point began moving east at 100 km/hr. The added vectors would give the object a resultant velocity of 1500 km/hr. This faster velocity would tend to make the object increase its radius of motion. However, gravity keeps objects near the Earth's surface from flying out into space, and the only way the object can increase its radius of motion is to bend its path toward the equator (clockwise). This deflection is described as the Coriolis effect.

Teaching Strategies

The concept of the Coriolis effect may be difficult for students to understand. You may need to perform various visual demonstrations with a globe to acquaint students with lines of latitude, the direction of the Earth's rotation, and the surface velocity of the Earth at different latitudes. As students work through the Exploration, remind them that Capt. Lease has a limited amount of fuel. Because fuel efficiency is important, students must recommend an appropriate average speed. Refer students to the CD-ROM articles to find out how speed and fuel are related.

As an extension of this Exploration, you may want to have students solve a similar situation that takes place in the Northern Hemisphere. Ask students how the Coriolis effect would influence a plane flying from the equator toward the North Pole. You may also want to explain why hurricanes rotate counterclockwise in the Northern Hemisphere and clockwise in the Southern Hemisphere, contrary to the behavior of other moving objects. Refer students to the CD-ROM articles to research this apparent contradiction.

Bibliography for Teachers

de Grasse Tyson, Neil. "The Coriolis Force." *Natural History,* 104 (3): March 1995, p. 76.

Lanken Dane, "Funnel Fury." *Canadian Geographic,* 116 (4): July/August 1996, p. 24.

Bibliography for Students

Gunther, Judith Anne. "Real-Time Weather." *Popular Science,* 246 (1): January 1995, p. 22.

Shepherdson, Nancy. "Controlling the Sky." *Boys' Life,* 85 (7): July 1995, p. 40.

Sweetman, Bill. "The Plane That Learns." *Popular Science,* 249 (1): July 1996, p. 40.

Other Media

Microsoft Flight Simulator 5.1 (for Macintosh and Windows 95)
CD-ROM or disk
Educational Resources
1550 Executive Dr.
Elgin, IL 60123
800-624-2926

Note: The above address is one of many locations from which you may order Microsoft products by mail. Microsoft products are also available at many retail stores. To locate the store nearest you, visit Microsoft's Web page at http://www.microsoft.com or call Microsoft's sales division at 800-426-9400.

Encourage students to search for relevant information about the Coriolis effect by exploring the Internet. Possible keywords include the following: *Coriolis effect, weather patterns, ocean currents,* and *hurricanes.* Students can also access information about *flying airplanes* and *flight paths* on the Internet.

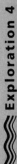

Latitude Attitude

1. Capt. Corey O. Lease needs to get supplies to a group of scientists on the Mertz Glacier. What has she asked you to do for her?

   _____

   _____   _____

   _____

   _____

2. In relation to Melbourne, where is the Mertz Glacier located?

   _____

   _____

3. How does the speed of the Earth's surface at the equator compare with the speed of the Earth's surface at the poles? (If you're not sure, check out the CD-ROM articles.)

   _____

   _____

   _____

   _____

   _____

4. How will you use the Accu-Flight Simulator to help Capt. Lease?

   _____

   _____

   _____

   _____

   _____

   _____

   _____

   _____

5. Use the Accu-Flight Simulator to determine the results of using different combinations of flight direction and average airspeed. Record your results in the table below.

Flight direction	Airspeed (km/hr)	Final location of the plane
130°	725	
140°	725	
150°	725	
180°	725	
210°	725	
220°	725	
230°	725	
130°	850	
140°	850	
150°	850	
180°	850	
210°	850	
220°	850	
230°	850	

6. What does the equipment on the lab's back counter demonstrate?

_____

_____

_____

_____

7. Describe how moving objects are affected by the Coriolis effect. (If you're not sure, check out the CD-ROM articles.)

_____

_____

_____

_____

_____

Record your answers in the fax to Capt. Lease.

Exploration 4

Name _____ Date _____ Class _____

FAX

To: Capt. Corey O. Lease (FAX 011-612-555-7996)

From:

Date:

Subject: Latitude Attitude

In which compass direction should I aim my plane?

180° (due south)	130°	140°	150°	210°	220°	230°

How fast should I fly the plane?

725 km/hr	850 km/hr

Please explain why it is necessary for me to fly the plane in the recommended direction in order to reach the Mertz Glacier.

_____

_____

_____

_____

_____

_____

_____

_____

_____

_____

_____

Latitude Attitude

The following articles can also be found by clicking the computer in the CD-ROM laboratory for Exploration 4:

- *The Coriolis Effect*
- *Wind, Water, and Weather*
- *What a Pilot Needs to Know*
- *Antarctica*

The Coriolis Effect

A Little Latitude

Lines of latitude are imaginary parallel lines that run east to west around the Earth. The equator, at a latitude of 0 degrees, divides the Earth in two halves, called the Northern Hemisphere and the Southern Hemisphere.

What Is the Coriolis Effect?

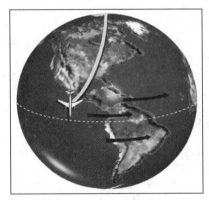

Because of the Coriolis effect, objects traveling south in the Northern Hemisphere appear to curve to the right (clockwise).

In 1835, Gustave-Gaspard de Coriolis, a French mathematician, explained that an object moving in a straight line perpendicular to the Earth's equator will appear to curve. This effect, called the **Coriolis effect,** is the apparent bending, or deflecting, of the path of a moving object due to the Earth's rotation. The Coriolis effect influences the paths of moving objects, such as airplanes, winds, and weather systems. In the Northern Hemisphere, objects moving north or south seem to veer to their right (clockwise). In the Southern Hemisphere, moving objects tend to veer to their left (counterclockwise).

So what causes this curving effect? The circumference of the Earth at the equator is larger than the circumference of the Earth at 30°, 60°, or at the poles. That means that a point on the equator travels farther during one of Earth's rotations (one day) than does a point at 30°, 60°, or near one of the poles. As a result, points on the Earth's surface near the equator move faster (about 1670 km/hr) than do points at 30° (about 1400 km/hr), 60° (about 800 km/hr), or at one of the poles (effectively 0 km/hr).

You can see how the Coriolis effect works by slowly spinning a record turntable clockwise. As the turntable rotates, imagine what would happen if you traced a line from the outside of the turntable to the center. You would get a line that curves to the left. Even though you trace a straight line, the clockwise rotation of the turntable causes the line that you trace to appear as a curve. Likewise, even though an object moving above the Earth's surface may follow a straight path, we will actually see a curve. That's the Coriolis effect.

Wind, Water, and Weather

Wind Patterns

Generally, air flows from areas of high pressure to areas of low pressure. Differences in pressure cause air to circulate between the poles and the equator, creating wind patterns. As the Earth rotates, the Coriolis effect influences the northerly or southerly path of the wind. In the Northern Hemisphere, winds circulate clockwise, while winds circulate counterclockwise in the Southern Hemisphere.

Ocean Currents

Winds that blow across the ocean's surface transfer energy to the ocean water, resulting in strong surface currents. For example, the trade winds create currents that affect the movement of ocean water near the equator. These ocean currents, as well as other ocean currents, are

Exploration 4

deflected by the Coriolis effect so that the currents flow clockwise in the Northern Hemisphere and counterclockwise in the Southern Hemisphere.

The Coriolis Effect and Weather Patterns

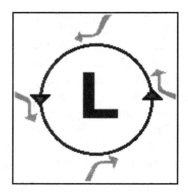

The movement of air masses causes most weather patterns, including thunderstorms, warm and cold fronts, and hurricanes. The Coriolis effect causes the deflection of air masses as air moves in the atmosphere. For example, in the Northern Hemisphere, parcels of high-pressure air moving toward a low-pressure system will be deflected to the right. This causes the low-pressure system to spin counterclockwise. For this reason, storm systems and hurricanes in the Northern Hemisphere spin counterclockwise, while in the Southern Hemisphere they spin clockwise.

What a Pilot Needs to Know

Speed and Fuel

One way to measure an airplane's speed is by ground speed. **Ground speed** is the speed of the plane with respect to the ground. Another measure is air speed. **Air speed** is the speed of the plane with respect to the air. If there is no wind (that is, if the air is not moving), then the air speed of the plane equals its ground speed. The air-speed indicator in the cockpit of a plane helps a pilot determine the plane's air speed.

The speed at which a plane flies affects the amount of fuel that the plane needs to reach its destination. Pilots calculate the amount of fuel necessary to reach a destination at ground speed, then they adjust these calculations to account for changes in speed due to winds.

Winds blowing in the direction that the plane is flying allow the pilot to fly farther with less fuel. On the other hand, winds that blow against the plane force a pilot to use more fuel to travel the same distance. Flying faster requires more power, and, therefore, more fuel.

Flight Path

An aeronautical chart is very important in air navigation.

When you plan a road trip, you can look at a road map to determine in what direction you need to drive to reach various destinations. However, if you were to follow those same directions in an airplane, you might not end up in the place you expected. Why? When you travel in the air (especially over great distances), you must account for the Coriolis effect.

When a plane sits motionless on the runway, it still has a velocity because the Earth's surface is rotating west to east at a given velocity. When the plane takes off, it still has that velocity with respect to the runway. As the plane begins to travel north or south, the plane will have a different velocity with respect to the Earth's surface because the surface velocity of the Earth is different at different latitudes. As a result, the plane will have a different velocity with respect to the Earth's surface. An airplane flying north from the equator will seem to move east because the surface velocity of the Earth decreases as latitude increases (toward the North Pole). If the pilot follows a straight path north, the plane will end up east of its destination. An airplane flying south from the North Pole will seem to move west because the surface velocity of the Earth increases as latitude decreases (toward the equator). This apparent deflection is an example of the Coriolis effect. Pilots must compensate for the Coriolis effect when planning their flight paths.

Suppose an airplane leaves Quito, Ecuador, near the equator, and flies directly toward New Orleans, Louisiana, about 30°N latitude. The eastward velocity of the Earth's surface is approximately 1670 km/hr at Quito. As the plane heads toward New Orleans, it seems to have some eastward velocity because the Earth's surface does not rotate as fast near New Orleans as it does at Quito. So the plane is actually traveling a little bit east (from a perspective above the Earth's surface). This apparent change in velocity has an effect on where the plane ends up. If the plane maintained its initial flight direction, it would end up landing east of New Orleans. To compensate for this, a pilot would aim the plane slightly northwest to ensure that the plane reaches its destination of New Orleans.

So what does this mean in terms of planning a flight direction and flight path? It means that it is necessary to adjust for the Coriolis effect when planning the direction of flight. Luckily, pilots can use computers to calculate the necessary adjustments when creating flight paths.

Antarctica

The Frozen Continent

The continent of Antarctica, an area that measures approximately 13 million square km, is almost entirely covered by a thick sheet of ice. The ice sheet was created by the buildup of snow over millions of years, and it contains over 70 percent of the world's freshwater supply. As the ice covering Antarctica piles up, it turns into glaciers and ice rivers that flow to the sea. A vast current called the Antarctic Circumpolar Current moves the cold waters into the rest of the world's oceans.

Antarctica is surrounded by the Pacific, Atlantic, and Indian Oceans. The Antarctic landmass is divided by the Transantarctic Mountains, and its geological features include an active volcano named Mount Erebus and the geographic and magnetic south poles. The Mertz and Innis Glaciers extend into eastern Antarctica just north of the south magnetic pole.

Who Wants to Live in Antarctica?

The frozen land and frigid weather mean little life thrives on Antarctica.

The South Pole region has an annual mean temperature of –49°C, and Antarctica holds the world record for cold temperature, –89.2°C. Winter lasts from mid-March to mid-September. During this time, the continent is submerged in darkness. So who on Earth would want to live in Antarctica?

Despite the extreme cold, scientists from all over the world gather regularly at a number of Antarctic research stations. At the South Pole Station, for example, astronomers study galaxy and star formation while astrophysicists keep an eye on the ozone layer. The McMurdo Dry Valleys are another hot spot for research on Antarctica. This unusual expanse of ice-free land, with its mountain ranges, meltwater streams, and arid terrain, draws botanists, geochemists, biologists, ecologists, and other scientists.

Exploration 4

Tunnel Vision

Key Concepts

Light bulbs connected in a parallel circuit are brighter than the same bulbs connected in a series circuit. Too much current flowing through a circuit can result in a burnout of the output device(s).

Summary

Seymore Rhodes, an aspiring inventor and bicycling enthusiast, is working on an electrical circuit for his bicycle helmet that will make his ride to school a safer one. Seymore has to ride his mountain bike through a dark tunnel on the way to school, and he is worried that drivers may not see him while he is riding through the tunnel. He has designed a circuit with three red lights that can attach to his bike helmet. However, the lights are too dim. Seymore needs to know how to make the lights brighter while keeping the circuit lightweight.

Mission

Advise a young inventor about the best circuit to use on his bicycle helmet.

Solution

Connecting the three bulbs in a parallel circuit powered by one battery is the best combination for Seymore's circuit. With this combination, the bulbs are very bright and the circuit is still lightweight.

Background

A typical cable used to wire houses for electricity is composed of three wires enclosed in a plastic casing. Two of the wires, one covered with gray insulation and one covered with blue insulation, carry alternating current to and from the outlets. The third wire is not insulated and acts as a safety feature. It is usually connected to a copper bar driven into the ground outside the house. If there is leakage of current, the electricity flows to the ground, preventing electrocution. Average homes are wired with four, five, or more parallel circuits. Several outlets are connected to each circuit. If too many appliances on one circuit are turned on at the same time, too much electric current moves through the wire. This can cause an overload, and that circuit will shut down, cutting off the electricity.

Fuses are designed to protect against overloads. A fuse is a safety device containing a short strip of metal with a low melting point. If too much current passes through, the metal melts, causing the fuse to "blow." This causes a break in the circuit, and the current no longer flows. Another device that protects circuits from overloads is a circuit breaker. One type of circuit breaker is a switch attached to a bimetallic strip. When the metal gets hot, it bends. This opens, or "trips," the circuit. This action does not harm the circuit breaker. After the overload has been corrected, the circuit breaker can be reset.

Teaching Strategies

Students may need extra time to become familiar with the circuit board in this Exploration. Explain to them that they will create circuits by closing switches, not by arranging wires. You may want to demonstrate how to create several different circuits. Students may also benefit from drawing circuit diagrams so that they can determine what types of circuits they create. In addition, students can create numerous circuits that contain only one or two light bulbs, but these will not qualify as possibilities for Seymore's circuit. In order to increase students' efficiency, you may want to suggest at the outset that students focus on creating circuits using all three light bulbs. As students begin to create circuits in the lab, remind them to note not only whether the bulbs light up but also how bright they are.

As an extension of this Exploration, you may wish to create similar electric circuits in your classroom. This may help students visualize what they are doing in Dr. Labcoat's lab. Using flashlight bulbs, batteries, and wires with alligator clips, they can easily construct the different circuits from the Exploration. If students construct the circuits themselves, remind them to be careful about giving too much voltage to circuits. Too much voltage can burn out the bulbs.

Bibliography for Teachers

Lipkin, Richard. "Teeny-Weeny Transistors." *Science News,* 147 (18): May 6, 1995, p. 287.

Steiger, Walter R., and Suk R. Hwang. "A Series-Parallel Demonstration." *The Physics Teacher,* 33 (9): December 1995, p. 590.

Bibliography for Students

Humberstone, E. *Electricity and Magnetism.* Usborne/EDC, 1994.

Kempter, Joseph. "Power to Go." *Astronomy,* 24 (1): January 1996, pp. 86–87.

Reis, Ronald A. "Get Charged." *Boys' Life,* 86 (1): January 1996, p. 53.

Other Media

Learning All About Electricity and Magnetism
CD-ROM for MAC or MS-DOS
Queue, Inc.
338 Commerce Drive
Fairfield, CT 06432
203-335-0906
800-232-2224

In addition to the above CD-ROM, students may also find relevant information about electricity and magnetism by exploring the Internet. Interested students can search for articles using keywords such as the following: *electricity, magnetism, electric current, electric circuits,* and *Ohm's law.*

Tunnel Vision

1. Seymore Rhodes wants to make his mountain-bike ride to school a safer one. What advice does he need from you?

_____

_____

_____

_____

2. What are the necessary components of an electric circuit? (If you're not sure, check out the CD-ROM articles.)

_____

_____

_____

_____

_____

_____

3. Look at the diagram that Seymore drew of his circuit. What kind of circuit is it?

_____

4. How is a series circuit different from a parallel circuit? (Hint: Check out the equipment on the lab's back counter.)

_____

_____

_____

_____

_____

_____

_____

5. Examine the equipment on the lab's front table. How will you use the switches and the batteries to help you answer Seymore's questions?

_____

_____

_____

_____

_____

_____

6. Create all of the necessary circuits you need to answer Seymore's questions. In the table below, record the switches you close, the number of batteries you use, and the type of circuit you create. Use the fourth column to record your observations about the light bulbs.

Switches closed	Number of batteries	Type of circuit	Effect on light bulbs

Exploration 5 Worksheet, continued

7. What effect does the number of batteries have on your circuits?

_____

_____

_____

_____

_____

_____

_____

_____

_____

8. What happens when too much current flows through a circuit?

_____

_____

_____

_____

_____

9. What is the relationship between current, voltage, and resistance? (If you're not sure, check out the CD-ROM articles.)

_____

_____

_____

_____

_____

_____

_____

Record your answers in the fax to Seymore Rhodes.

Name _____ Date _____ Class _____

FAX

To: Seymore Rhodes (FAX 406-555-7209)

From:

Date:

Subject: Tunnel Vision

How many batteries should I use?

1	2

Please describe the best way to wire the lights for my helmet.

_____

_____

_____

_____

_____

_____

_____

_____

·❀·

For Internal Use Only

Please answer the following questions for my laboratory records. Scientists must always keep good records. Dr. Crystal Labcoat

What is the best type of circuit for Seymore's lights?

SERIES	SERIES-PARALLEL	PARALLEL

Name _____ Date _____ Class _____

Describe the changes that you made to Seymore's original circuit.

_____
_____
_____
_____
_____
_____
_____
_____
_____
_____
_____
_____
_____
_____
_____
_____
_____
_____
_____
_____
_____
_____
_____

Tunnel Vision

The following articles can also be found by clicking the computer in the CD-ROM laboratory for Exploration 5:

• *The Essentials of Electricity*
• *Circuits—Making the Connection*

The Essentials of Electricity

What Is Electric Current?

A bolt of lightning can be described as a big electric spark in the sky. More specifically, it is an electric current passing from one cloud to another cloud or to the Earth, or from the Earth to a cloud. An **electric current** is the flow of electrons from an object that has many electrons to another object that has too few electrons. The uneven distribution of electrons in the sky is a result of friction between the moving clouds and the surrounding air. To understand this process, think about what happens when you scuff your feet on a carpet and reach for a doorknob. Ouch! The shock you experience is the flow of electrons, or electric current, from your hand to the doorknob.

Although these examples of electric current occur naturally, they are not useful in our everyday lives. To make electric current useful, an electric circuit is required. An **electric circuit** is a continuous pathway for an electric current. The components of a simple circuit include a source of electricity, such as a battery; an output device, such as a light bulb; and wires. Many circuits also include switches, which allow the circuit to be turned on and off. When the switch is off, the pathway of the circuit is interrupted. This situation is called an **open circuit,** and no current can flow. When the switch is turned on, the circuit is complete and the current flows. This situation is called a **closed circuit.**

How Does a Flashlight Work?

A flashlight is an example of a simple circuit. A flashlight generally consists of a plastic case, two batteries, a flashlight bulb, a switch, and a metal strip. When the flashlight is off, the switch is open, and the metal strip does not touch the flashlight bulb. When you turn the flashlight on, the switch closes the circuit by moving the metal strip so that it touches the flashlight bulb. Electrons then flow from the negative end of the battery, through the metal strip, through the filament in the bulb, and then to the positive end of the battery. As the current flows through the filament, the filament heats up and glows to produce light.

Circuits—Making the Connection

Circuit Symbols

Drawing a diagram is a useful way to describe a circuit. Remember that circuits can involve various output devices, such as light bulbs, toasters, hair dryers, microwave ovens, and almost anything else you can think of that you plug in. In the following sections, we will discuss various kinds of circuits, but we will use light bulbs in every example.

Series Circuits

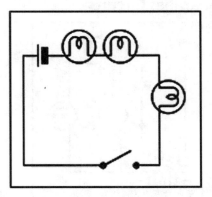

In a series circuit, there is only one pathway for the current to follow. If there is a break in the circuit, the flow of current is disrupted, and none of the bulbs light. The fact that the bulbs are connected in series also affects the brightness of the bulbs. Because the light bulbs share the same current, three bulbs in series are dimmer than one single bulb connected to the battery.

Parallel Circuits

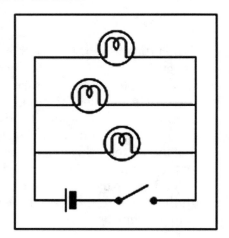

In a parallel circuit, current is divided into two or more branches. When current reaches a branching point, some current goes along one branch and some goes to another branch. If a break occurs in one branch, current can continue to flow through other branches and the bulbs along those branches stay lit. The fact that the bulbs are connected in parallel also determines the brightness of the bulbs. The battery powers each bulb individually, so each bulb emits its standard brightness. In other words, three bulbs in parallel are just as bright as one single bulb connected to the battery.

Series-Parallel Circuits

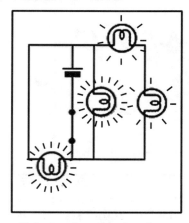

Series-parallel circuits contain elements of both series and parallel circuits.

Path of Least Resistance

Electric current tends to follow the path of least resistance. **Resistance** can be thought of as electrical friction—it is the opposition to current in a circuit. All output devices (such as light bulbs, etc.) offer some resistance to electric current. In a circuit that involves light bulbs, electric current tends to follow the path with the fewest bulbs or with bulbs with the least resistance. Keep in mind that some of the current flows through each complete pathway. How much current flows through a particular pathway depends on the resistance of that pathway as compared to the other pathways. In a series circuit, current follows only one pathway, so each bulb lights equally if each bulb provides the same resistance. In a parallel circuit, however, more current always goes into the branches of the circuit with lower resistance, and less current goes into the branches with higher resistance.

Resistance can be calculated using Ohm's law, which describes the relationship between voltage, current, and resistance in an electric circuit. Ohm's law states:

$$\text{current (I)} = \frac{\text{voltage (V)}}{\text{resistance (R)}}$$

$$\text{or } I = \frac{V}{R}$$

You can use the equation to solve for any one of the variables. To find the resistance in a circuit, you would divide V by I. The ratio V/I demonstrates that, for a given resistance, current and voltage are directly proportional. This allows you to predict the effect of changes in the circuit. Doubling the voltage in a circuit will double the current in the circuit.

Circuit Failure

Although circuits are easy to design and construct, sometimes problems arise that cause the circuit to fail or to not operate properly. For example, if too much current is introduced into a circuit involving a light bulb, the bulb can burn out. Remember that Ohm's law states that current and voltage are directly proportional. That means that for a circuit with a given resistance, an increase in voltage means an increase in current. In our example of the flashlight, that means that too much voltage from the batteries can burn out the bulb.

Another example of circuit failure is a short circuit. A **short circuit** occurs when the bulbs and wires are connected in such a way that the current bypasses the bulb. When the current takes such a short cut, the bulb won't light.

Sound Bite!

Key Concepts	Sound waves can be described by their wavelength, frequency, and phase. Active noise control uses deconstructive interference to reduce the amplitude of a sound wave.
Summary	Mr. Cy Lintz, the owner of a pet store, is losing business because his usually mild-mannered guinea pigs have been biting the customers! He suspects that the cause of this biting outbreak is the constant hum of freezers coming from the new ice cream shop next door. Mr. Lintz knows that guinea pigs become highly agitated when exposed to loud noises, and he believes this is why the guinea pigs are displaying this aggressive behavior. Mr. Lintz has read about a new technology used to eliminate noise pollution and would like to know how it might be applied to solve his problem.
Mission	Apply the concept of active noise control in an attempt to soothe some agitated guinea pigs.
Solution	A sound in phase 2 with a frequency of 225 Hz and a low amplitude will eliminate the offending sound. Its waves destructively interfere with those of the offending sound (which is in phase 1 and has a frequency of 225 Hz and a low amplitude).
Background	Your ear is specialized for picking up sound waves and transmitting them to your brain. A sound, such as a telephone ring, causes particles in the air around the phone to vibrate. These vibrations, or sound waves, travel through the air in longitudinal waves. The outer ear, or pinna, catches the sound waves and directs them into the ear canal. When the sound waves reach the eardrum at the end of the canal, the eardrum begins to vibrate. The vibrations are passed along to three bones, the hammer, anvil, and stirrup, on the other side of the eardrum. These bones act like levers, multiplying the force of the vibrations. When the stirrup vibrates, the energy of the vibrations passes through the oval window into the cochlea, where it causes waves in the fluid there. Along the length of the cochlea are tiny hair cells that vibrate in response to the frequency of the waves. Nerve cells at the base of the hair cells change the vibrations into electrical impulses. These impulses are carried along the auditory nerve to the brain, where they are interpreted as the ring of a telephone.

Teaching Strategies

Although this Exploration requires students to analyze aspects of sound by examining transverse waves on oscilloscope screens, you may need to emphasize that sound waves are actually longitudinal waves. You may refer students to the wall chart and discuss how oscilloscopes convert longitudinal waves to transverse waves. Converting longitudinal waves to transverse waves on an oscilloscope screen allows you to "see" the frequency and amplitude of a sound. In addition, you may want to explain the function of the two phase buttons in the Exploration. They do not represent the only possible positions for a transverse wave. Refer students to the CD-ROM articles for a discussion of waves that are in phase or out of phase. Explain that the phase options in this Exploration represent an opportunity to put the waves exactly in phase or exactly out of phase.

As an extension of this Exploration, you may wish to conduct an investigative discussion of how sounds travel through different mediums. In this Exploration, the offending sound traveled through a connecting vent between the pet store and the ice cream shop. You could determine how the situation would be different if the humming noise were instead passing through an adjoining wall, the ceiling, or the floor. *(The sound could be harder to hear, more muffled, and more difficult to analyze in terms of its frequency and amplitude, for example.)* Ask students how such a change would affect how Mr. Lintz could implement active noise control. *(He couldn't simply generate a sound through the vent, for example.)*

Bibliography for Teachers

Sungar, Nilgun. "Teaching the Superposition of Waves." *The Physics Teacher,* 34 (3): April 1996, p. 236.

Wolkomir, Richard. "Decibel by Decibel, Reducing the Din to a Very Dull Roar." *Smithsonian,* 26 (11): February 1996, pp. 56–65.

Bibliography for Students

Foster, Edward J. "Switched On Silence." *Popular Science,* 245 (1): July 1994, p. 33.

Johnson, Julie. "Beating Bedlam." *New Scientist,* 144 (1947): October 15, 1994, pp. 44–47.

Other Media

Standing Waves and the Principle of Superposition
Video
Encyclopædia Britannica Educational Corporation
310 S. Michigan Ave.
Chicago, IL 60604-9839
800-554-9862

In addition to the above video, students may find relevant information about sound and sound waves by exploring the Internet. Interested students can search for articles using keywords such as *sound, sound waves, constructive interference, deconstructive interference,* and *acoustics.* Students can also access information about *active noise control* on the Internet.

Sound Bite!

1. Mr. Lintz is worried about the aggressive behavior of his guinea pigs. Describe his problem and what he has asked you to do.

_____

_____

_____

_____

_____

2. Examine the equipment on the back counter of the lab, and then describe what sound waves are.

_____

_____

_____

_____

_____

_____

_____

3. What purpose does an oscilloscope serve? (Hint: Check out the wall chart.)

_____

_____

_____

_____

4. What is destructive interference? (If you're not sure, check out the CD-ROM articles.)

_____

_____

_____

_____

_____

_____

Name _____ Date _____ Class _____

5. Conduct your experiment using the lab equipment. In the table below, record the frequency, amplitude, and phase you selected for the generated sound. Use the fourth column to describe the wave you see in the center oscilloscope screen when the generated sound is combined with the offending sound.

Frequency (Hz)	Amplitude	Phase	Results

6. Based on your results, what are the offending sound's frequency, amplitude, and phase?

_____

_____

_____

_____

7. What does *phase* refer to? (If you're not sure, check out the CD-ROM articles.)

_____

_____

_____

_____

_____

Record your answers in the fax to Mr. Lintz.

Exploration 6

Name _____ Date _____ Class _____

FAX

To: Mr. Cy Lintz (FAX 707-555-8988)

From:

Date:

Subject: Sound Bite!

Which settings eliminate Mr. Lintz's noise pollution?

Frequency (hertz)	175	225	275	325
Amplitude	low	medium	high	
Phase	1	2		

·❊·

For Internal Use Only

Please answer the following questions for my laboratory records. Scientists must always keep good records. Dr. Crystal Labcoat

Explain how sound travels.

_____

_____

_____

_____

_____

_____

_____

_____

_____

Name ____________________ Date __________ Class __________

Exploration 6 Fax Form, continued

Explain how active noise control is used to eliminate noise pollution.

Exploration 6

Sound Bite!

The following articles can also be found by clicking the computer in the CD-ROM laboratory for Exploration 6:

- *Makin' Waves*
- *Wave Basics*
- *Sound Advice*

Makin' Waves

Good Vibrations

All sound is produced by vibrations. In the case of a bumblebee, you can see the vibrations in the movement of its wings. You can also see the vibrations of a guitar string that has been plucked. Most sound, however, is produced by vibrations that are too small and too fast to see with the unaided eye.

Sound begins when a vibrating object pushes on the particles around it. The particles near the object become more dense (more tightly packed) than the other particles around them. The dense particles then spread apart, pushing against other particles. This causes the other particles to crowd together and then spread apart in a back-and-forth motion. Particles continue to crowd together and spread apart in succession as the sound travels away from its source. The result is waves of sound.

Sound can travel through different mediums. Usually, the sounds that we hear are caused by sound waves that travel through the air, but we can also hear sound that travels through liquids and solids. Have you ever listened to sounds underwater in a swimming pool? While underwater, you can hear the sounds of shouts and splashes as sound waves pass through the water. Sound can also pass through solid materials. For example, you can hear voices through a wooden door. It is impossible for sound to travel where there are no particles to vibrate, so sound cannot travel through space or through a vacuum.

Transverse Waves

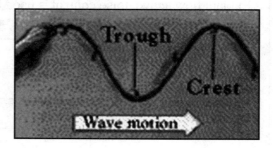

A **wave** is a disturbance that transmits energy as it travels through matter or space. A **transverse wave** is a wave in which the motion of the particles of the medium is perpendicular to the path of the wave. A transverse wave has a **crest,** the uppermost part of the wave, and a **trough,** the lowest part of the wave. You can generate a transverse wave by holding the end of a rope in your hand and moving your hand up and down. The energy moves along the wave, but the medium itself does not move. In other words, as you move the rope up and down, the rope itself does not travel horizontally, but you do send energy along the length of the rope.

Longitudinal Waves

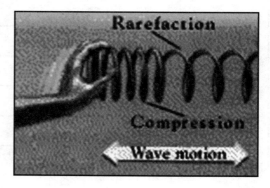

Not all waves move up and down like transverse waves. **Longitudinal waves** move back and forth, parallel to the direction of motion of the wave. Unlike transverse waves, longitudinal waves do not have crests or troughs. Instead, they have regions of **compression,** in which particles are pushed close together, and regions of **rarefaction,** in which particles are pulled apart from each other. Longitudinal waves consist of alternating compressions and rarefactions moving through a medium. If you alternately compress and stretch a spring, you can observe longitudinal waves. Notice again that the particles of the spring are not carried along with the compressions and rarefactions of the wave; they pass the energy of the wave along to the neighboring particles and return to their original position.

Wave Basics

Amplitude

Amplitude refers to the height of the wave. For a transverse wave, the amplitude is half the vertical distance between a crest and a trough. For a longitudinal wave, the amplitude is represented by the relative amount of compression and rarefaction the wave experiences. For both transverse and longitudinal waves, the greater the energy that the wave carries, the greater the wave's amplitude.

Wavelength

Wavelength refers to the distance between two identical points on neighboring waves. You can measure wavelength by measuring the distance between two crests or two troughs of a transverse wave or by measuring the distance between two compressions or two rarefactions of a longitudinal wave.

Frequency

Frequency refers to the number of waves produced in a given time. Frequency is usually measured in Hertz (Hz), the number of vibrations per second. Something that vibrates at a rate of 100 Hz vibrates 100 times per second. You could measure frequency in a transverse wave by counting the number of crests or troughs that pass a point in a given amount of time. For a longitudinal wave, you could count the number of compressions or rarefactions that go by a certain point in a given amount of time.

Wave Speed

Wave speed refers to the speed at which a wave travels and is usually measured in meters per second. The speed of a wave can be measured by observing how fast a single compression or crest moves through a medium. The speed of sound in air is about 346 m/s. In general, sound waves travel even faster through liquids and solids because liquids and solids have less space between particles.

Phase

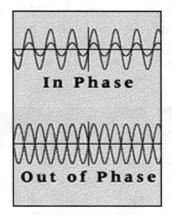

Phase is a term used to describe the positions of the crests and troughs of two individual waves relative to each other. For example, two waves of a given frequency are said to be in phase if the crests and troughs of one wave match up with the crests and troughs of the other wave. Two waves of a given frequency are said to be out of phase when the crests and troughs of the two waves do not match up.

Wave Interference

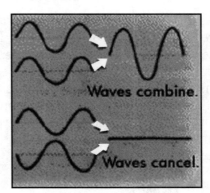

Waves that undergo constructive interference produce a single, stronger wave. Waves that undergo destructive interference tend to cancel each other out.

When you throw a pebble into a pond, you see circular ripples form in the water. But what happens when you throw more than one pebble into the pond at a time? Ripples from the pebbles run into one another and cause interference. When two crests or two troughs meet, the waves combine and form higher crests and deeper troughs. This is called **constructive interference,** and it causes an increase in wave amplitude. If a crest and trough of equal amplitude and frequency meet, they cancel each other out. This is called **destructive interference,** and it causes a decrease in amplitude. See "The Technology of Noise Control" to see how destructive interference can be used to combat unwanted sounds.

Sound Advice

How to Change a Sound

One way to change a sound is to increase or decrease the sound's amplitude. Recall that the amplitude of a sound is related to the amount of energy used to generate the sound waves. If you pluck a guitar string gently, the distance that the string travels in one vibration will be small compared with the distance the string travels in one vibration if you pluck the string forcefully. Changing the amplitude of the vibration changes the loudness of the sound produced by the guitar string. The same thing happens with your vocal chords. When you speak softly, your vocal chords vibrate with a small amplitude. When you speak louder, you increase the amplitude of vibration.

A change in frequency also alters sound. The fewer the vibrations, the lower the frequency and the lower the pitch. Increasing the number of vibrations increases the frequency and raises the pitch of the sound. When you sing low musical notes, your vocal chords are vibrating fewer times per second than they do when you sing high notes.

Turn That Down!

Noise pollution is sound that harms human health. High-energy sounds usually have high volume. **Volume,** or loudness, is measured in **decibels (dB).** Sound energy approximately doubles for every 3 dB increase in loudness. A normal conversation has a volume of about 60 dB, and a rock concert typically has a volume of about 120 dB. The 60 dB increase in loudness from the conversation to the rock concert corresponds to an increase in sound energy by a factor of 1,000,000. Sounds of 90 dB or greater cause permanent hearing damage if exposure to the sound is continuous, and sounds of 170 dB can cause total deafness.

Workers exposed to excessive noise—in factories and on airport runways, for example—must take precautions to protect their hearing. Often you see workers wearing large headphones that protect their ears from loud noises in the workplace. But you don't have to work on a runway to be at risk for hearing damage. Many sounds we encounter in our day-to-day lives can pose a threat to our hearing. Lowering the volume of the TV or your headphones, using ear plugs at rock concerts, and avoiding unnecessary exposure to loud machinery and traffic will help protect you from potential hearing loss.

The Technology of Noise Control

Because of the physical dangers caused by noise pollution, it is desirable to combat offending sounds. Have you ever looked at the ceiling in a busy restaurant? Often the ceiling tiles are specially designed to absorb as much sound as possible. **Passive noise control** is a method of absorbing sound waves or insulating an area from sound. Ear plugs are a common example of passive sound control. Most forms of passive noise control reduce medium and high frequency sounds but are less effective at reducing low frequency sounds.

Another way to combat noise pollution is by active noise control. **Active noise control** uses destructive interference to reduce the amplitude of a sound wave—you actually make more sound to control the offending sound. A speaker is set up to produce sound waves that are the mirror image of the unwanted noise. When the sound waves from the speaker meet the sound waves from the source of noise, the waves cancel each other out. The process works best at low frequencies and in close quarters, such as a small room or an air duct, or in headphones.

In the Spotlight

Key Concepts

A filter is a colored lens that absorbs certain wavelengths of light and transmits others. Different colors of light can be produced by mixing the primary colors of light.

Summary

Ms. Iris Kones is the director of a small community theater. She is preparing for the theater's first big production, which will require many different-colored spotlights. The theater is equipped with two spotlights and 12 colored filters. Ms. Kones wants to know how many different colors of light she can create using the available filters.

Mission

Enlighten the director of a community theater about how to create different colors of light.

Solution

Thirteen colors of light can be produced from the available filters. These colors are as follows: white, red, blue, yellow, green, magenta, cyan, burnt orange, lemon lime, medium blue, raspberry, vivid violet, and pale green.

Background

The lighting design is a crucial component of a stage production. Stage lights can establish the atmosphere or mood of a scene, illuminate key portions of the stage, emphasize dramatic action, and even produce special effects. Lights can also enhance the set of a production by projecting a scenic background onto a screen. Most productions require a range of lighting requirements. For example, one scene may take place outdoors at sunset and another scene may take place indoors at midday. Appropriate lighting is necessary to capture the essence of both situations.

Different kinds of lights serve different purposes. For example, spotlights are positioned above and behind the audience and direct a concentrated beam of light onto the stage. Background lights are positioned closer to the stage and provide overall color and light to the entire stage. Lights used for background lighting are often a series of lights attached to a beam above the stage. Each light on the beam can be aimed at a different area of the stage to ensure the best coverage. Other pieces of lighting equipment also enhance the lighting design. For example, colored filters can be used to create different colors of light, and various lenses can intensify or soften the brightness of a light. Most stage lights are controlled by a lighting board that has dimmer switches; these switches can be preset at specific levels for the appropriate scenes. Most lighting boards are computerized so that the board adjusts automatically with the push of a button.

Exploration 7

Teaching Strategies

Encourage students to work efficiently through this Exploration. Students can reduce time spent in the lab by detecting color combinations quickly and by noting that testing each filter on one spotlight with every filter on the other spotlight is not necessary. For example, red and cyan light will produce white light no matter which spotlight is red and which is cyan. You may wish to explain to students what complementary colors are. The complementary color of any primary color of light is the color formed by combining the two other primary colors. The complement of blue is yellow (red light plus green light); the complement of red is cyan (blue light plus green light); and the complement of green light is magenta (red light plus blue light). Two complementary colors produce white light when mixed. For example, green light plus magenta light produces white light.

As an extension of this Exploration, you may want to conduct some activities that demonstrate color-vision effects, such as successive contrast. Cut out various shapes from colored construction paper. Have students stare at the colored images for about 30 seconds, and then have them look at a white surface. They will see an afterimage that has the same shape as the original image but with a different color. For example, if the original image was red, the afterimage will be green; and if the original image was blue, the afterimage will be yellow.

Bibliography for Teachers

Peterson, Ivars. "Butterfly Blue." *Science News,* 378(19): November 4, 1995, pp. 296–297.

Winters, Jeffrey. "Vanishing Electrons." *Discover,* 17(7): July 1996, p. 38.

Bibliography for Students

"Electronic Seeing Eye." *Popular Mechanics,* 176(6): June 1996, p. 16.

Bulla, Clyde Robert. *What Makes a Shadow?* Harper Collins, 1995.

Other Media

Learning All About Light & Lasers
CD-ROM for MAC or MS-DOS
Queue, Inc.
338 Commerce Drive
Fairfield, CT 06432
203-335-0906
800-232-2224

In addition to the above CD-ROM, students may find relevant information about light and color by exploring the Internet. Interested students can search for articles using keywords such as *colors of light, visible spectrum,* and *lasers.* Students may also find information about *stage lighting* and *lighting design* on the Internet.

In the Spotlight

1. Ms. Kones is a little in the dark about the lighting design for her first stage production. What questions do you need to answer for her?

_____

_____

_____

_____

_____

_____

2. Describe the equipment that Dr. Labcoat has set up on the front table.

_____

_____

_____

_____

3. How do colored filters work to create different colors of light? (If you aren't sure, check out the CD-ROM articles.)

_____

_____

_____

_____

_____

_____

_____

_____

Exploration 7

Exploration 7 Worksheet, continued

4. As you try different combinations of filters, record your results, including the color equations for each color, in the table below.

Filter 1	Filter 2	Resulting color	Color equation

Exploration 7 Worksheet, continued

---

5. What are the primary colors of light?

_____

_____

6. What are the secondary colors of light?

_____

_____

7. What happens when you mix two primary colors of light?

_____

_____

_____

8. What happens when you mix two secondary colors of light?

_____

_____

_____

9. What differences do you notice between mixing the colors of paint on the back counter and mixing colors of light? (If you aren't sure, check out the CD-ROM articles.)

_____

_____

_____

_____

_____

_____

_____

_____

_____

Record your answers in the fax to Ms. Kones.

Exploration 7

FAX

To:	Ms. Iris Kones (FAX 409-555-2017)
From:	
Date:	
Subject:	In the Spotlight

How many different colors of light (including white light) can be produced using Ms. Kones's two spotlights and twelve filters? _____

Write the color equation for each color that you produced.

_____

_____

_____

_____

_____

_____

_____

_____

_____

_____

_____

_____

_____

_____

_____

_____

Exploration 7 Fax Form, continued

Describe how colored filters produce different colors of light.

_____

_____

_____

_____

_____

_____

_____

Please explain the difference between mixing colors of paint and mixing colors of light.

_____

_____

_____

_____

_____

_____

_____

_____

_____

_____

Exploration 7

In the Spotlight

The following articles can also be found by clicking the computer in the CD-ROM laboratory for Exploration 7:

- *The Nature of Light*
- *Color Me Beautiful*
- *Lighten Up!*

The Nature of Light

The Particle Theory

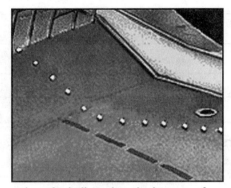

When the ball reaches the bottom of the hill, it changes direction.

Isaac Newton, who is famous for his work with motion, was one of the first scientists to propose the theory that light is composed of particles. He noticed that different colors of light could be produced by shining white light through a prism. The different colors are a result of **refraction,** or the bending of the light as it goes through the prism. Newton argued that this showed that light consists of a stream of particles because the paths of particles also bend when passing through different materials. Consider a ball rolling down a hill. When the ball reaches the bottom of the hill, its path bends, or refracts, and the ball changes direction. Newton suggested that because light also changes its direction, it must be composed of particles.

Newton also noticed that light travels in a straight line, another fact that indicates that light is a stream of particles. An object placed in the path of light casts a shadow; in other words, the object stops the light. In the same way, a barrier placed in the path of a rolling ball stops the ball's motion.

The Wave Theory

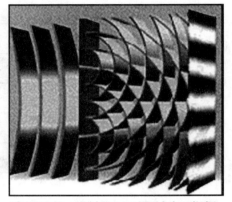

Thomas Young demonstrated that light can behave like a wave.

About the same time that Newton proposed his particle theory, a wave theory of light was proposed. The wave theory was not initially accepted because light could not be shown to diffract. **Diffraction** is the bending of waves around barriers. We know that sound waves diffract because we can hear sounds around a corner or outside a doorway. However, we cannot see around a corner, so it does not seem possible that light could be a wave.

However, in 1801, a scientist named Thomas Young conducted an experiment that proved that light does diffract and therefore behaves like a wave. Young made two narrow, parallel slits in a card. He placed a light source on one side of the card and a screen on the other. Young reasoned that if light consists of waves, the light would diffract as it passed through the slits. Young's hypothesis proved true. In this experiment, each slit acts as a separate source of light waves, and the waves coming through the slits **interfere** with each other; that is, crests coincide with crests and troughs coincide with troughs (constructive interference), resulting in bands of light on the screen. When crests coincide with troughs (destructive interference), they cancel each other out and dark bands appear on the screen.

The Particle-Wave Theory

Today scientists realize that both the wave model and the particle model are needed to describe all the behaviors of light. This has led to the particle-wave theory of light. How can light exhibit the qualities of both particles and waves? Perhaps a third explanation—the true nature of light—awaits discovery.

Electromagnetic Radiation

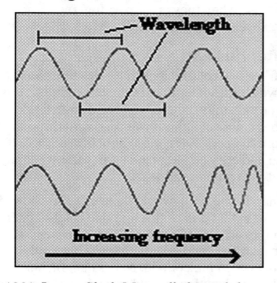

In 1864, James Clerk Maxwell showed that light consists of electromagnetic waves. Electromagnetic waves are waves that carry both electrical and magnetic energy and move through a vacuum at the speed of light (300,000,000 m/s). Heinrich Hertz continued Maxwell's work and showed that electromagnetic waves have a wide range of wavelengths ranging from thousands of kilometers to billionths of a centimeter. Together, these waves make up a type of energy called **electromagnetic radiation,** and they are represented by what we call the electromagnetic spectrum.

When you think of light, you probably think of sunlight, light from lamps, or different colors of light. But these kinds of light are only a part of a much larger phenomenon called the electromagnetic spectrum. The **electromagnetic spectrum** consists of waves of electromagnetic radiation that vary in wavelength and frequency. **Wavelength** is the distance between two corresponding points in a wave, such as from one crest to the next. **Frequency** refers to the number of wavelengths that pass a point during one second.

The Electromagnetic Spectrum

Light that we can see, or **visible light,** is only one part of the electromagnetic spectrum. Visible light contains electromagnetic waves of wavelengths that range from about 400 to 760 nanometers (billionths of a meter). Blue light, for example, has a shorter wavelength and a higher frequency than red light does. When all of the frequencies of visible light are put together, the light appears white.

The types of electromagnetic radiation that we cannot see in the electromagnetic spectrum include radio waves, X rays, infrared light, and ultraviolet light. These types of electromagnetic radiation consist of wavelengths that are either too small or too large for our eyes to see.

Color Me Beautiful

Prisms and Colors of Light

Visible light is only a small portion of the entire range of the electromagnetic spectrum. The color of visible light varies according to its frequency. All the colors of visible light combine to form white light. With a prism, we can see that the process of refraction separates white light into its component colors. Light of shorter wavelengths, such as blue light, bends more than light of longer wavelengths, such as red light. When white light separates, the colors always appear in the same order, from longest wavelength to shortest wavelength. This separation of white light into different colors of light also occurs in nature. You can see it happening every time you see a rainbow. A rainbow results when water vapor functions like a prism and bends the white light from the sun.

Colors of Objects

What makes a red apple red? Why can you say a blue shirt is actually every color *but* blue? Objects appear to be certain colors because white light is shining on the objects and the objects are absorbing, transmitting, or reflecting specific portions of visible light. The color of an object is due to the color of light that the object reflects. For example, if all frequencies of visible light are reflected from an object, the object appears white. If all frequencies of light are absorbed, the object appears black. A red apple

Exploration 7

looks red because it reflects red light and absorbs all light that is not red. You can say that a blue shirt is every color *but* blue because it absorbs every color of light but blue.

Pigments

If you were to add a drop of red food coloring to a glass of milk, the resulting mixture would be pink. The food coloring contains red pigments that add color to the milk. **Pigments** are basically colored substances that have been processed so they can be used to color other substances. A blue crayon is blue because blue pigment has been added to the crayon wax. The blue pigment reflects only blue light. Colored paint is also a type of pigment.

Primary Colors

We call certain colors primary because all other colors can be made from these colors. When we talk about primary colors, we usually specify whether we are referring to the **primary colors of light** or the **primary colors of paint.** The primary colors of light are red, green, and blue. When you mix these colors together, you get white light. By mixing the primary colors of light in different combinations, you can create all other colors of light, including cyan, magenta, and yellow.

The primary colors of paint are red, yellow, and blue. All other colors of paint can be made from these primary colors. Artists can have an incredibly colorful palette just by combining these three colors. Think about what happens when you mix blue, red, and green paint together. Do you get the same result as when you mix these colors of light (white)? No, because the primary colors of light are different from the primary colors of paint. You can find out why by reading the next section.

Color Formation

When you mix various colors of light, you are adding different wavelengths of light together and "building" white light. This is called **additive color formation.** White light is the presence of all colors of light. Blackness is the absence of all colors of light.

Mixing colors of paint is different from mixing colors of light. Suppose you have a white piece of paper. That piece of paper is reflecting all colors

of light. If you paint a splotch of red paint on the paper, the red paint will absorb all the colors of white light except for red light, which it reflects. If you mix paint of another color with the red paint, the resulting blend absorbs more colors of light from white light than either paint would absorb separately. Thus, mixing paint is called **subtractive color formation** because it tends to subtract colors of light from white light. Red, blue, and yellow are the primary colors of paint because when they are mixed together, black, the absence of light, results.

Filtering Light

A **filter** is a colored lens that absorbs certain wavelengths of light and transmits others. A red filter transmits only red light and absorbs all other colors of light. If you pass light through more than one filter, more light is absorbed. If you add enough filters, no light is transmitted and the result is darkness.

What do you think would happen if you allowed the light that passes through one colored filter to overlap on a screen with light that passes through a second colored filter? This would allow you to mix, or add, colors of light, and new color combinations would arise. Using filters that are the primary colors of light (red, blue, and green) allows you to create all other colors of light.

Lighten Up!

Holograms

Holograms are images that are made using laser light. In a holographic camera, a laser beam is split in two. One beam strikes the subject of the image and is reflected onto a piece of film. The second beam hits the film without striking the subject. The two beams interact to produce a microscopic interference pattern that is recorded on the film. In the proper lighting, the image produced by the film appears three-dimensional.

Lighting Systems

Creating the lighting for concerts, plays, and movie sets requires a solid understanding of both additive and subtractive color formation. Light designers use a variety of colored filters to create their spotlight designs. Light designers also consider the color of the backdrop and stage. If the backdrop is colored and the designer uses colored light, subtractive color formation will affect what the audience sees. For example, a blue light will make red, yellow, green, and orange objects look black because they reflect only red, yellow, green, and orange light, respectively, and absorb all other colors of light, including blue.

Exploration 7

DNA Pawprints

Key Concepts

DNA is the chemical in chromosomes that carries the genetic information that instructs the cells of living organisms. DNA can be manipulated to produce a DNA fingerprint, which can be used for identification.

Summary

Ms. Jean Poole breeds and shows border collies. She wants to enter three of her younger dogs in an upcoming show, but she lacks the necessary information to complete the dogs' pedigrees. She cannot remember which of her older male dogs sired each of her younger dogs. Ms. Poole has sent blood samples from the young dogs, their mother, and their possible fathers to Dr. Labcoat. Ms. Poole needs to know which older male fathered which young dog, and she would also like some information about the test used to determine their fathers.

Mission

Help a dog breeder obtain the necessary information for her dogs' pedigrees.

Solution

King sired Sugar and Merlin, and Roy sired Domino. The process used to determine the heredity of the young dogs is called DNA fingerprinting.

Background

Puppies of a litter born of purebred dogs share many traits, from size and coloring to temperament. At dog shows, breeders pay careful attention to the parents and offspring of prize-winning animals. Top dog breeders will breed prize-winning parents in the hopes of producing prize-winning offspring. Successive breeding does not always produce desirable results. For example, about 30 percent of all purebred Rottweilers inherit a recessive trait that causes a condition known as *hip dysplasia,* a deformity of the ball-and-socket joint where the hip bone meets the leg bone. Even a mild case of hip dysplasia disqualifies a dog from competition. In the worst cases, it leaves a dog crippled and in pain. Only an artificial hip can eliminate the problem.

Competitive breeders are aware of the high rate of hip dysplasia among purebreds. Dogs that have the deformity are prevented from breeding. Why, then, are there still so many puppies born with the disorder? First, many dogs that carry the recessive allele that causes the disorder do not show outward signs of hip dysplasia. When two of these carrier dogs breed, some of their puppies can receive two copies of the recessive allele. These puppies inherit the trait for hip dysplasia. Another factor that contributes to the widespread inheritance of hip dysplasia is *inbreeding.* Related prize-winning dogs are frequently bred together, and their puppies are often sold to other breeders. The genetic material from just a few parent dogs recombines again and again, and the chance of two dogs carrying the same recessive gene increases as offspring from closely related litters breed. Therefore, the probability that offspring will inherit both recessive alleles increases.

Teaching Strategies

As students work through this Exploration, remind them that although they are comparing DNA samples by comparing bands of DNA fingerprints, DNA is not actually a strand of bands. Refer students to Transparency #3 on the back counter in the lab for an actual depiction of the shape of DNA. You may also wish to reinforce among students the idea that a living organism that reproduces sexually receives half its genetic information from one parent and the other half from the other parent. This may help students understand why the DNA fingerprints of each young dog display half Bella's bands and half the appropriate father's bands.

As an extension of this Exploration, you may encourage students to compare the DNA fingerprints of Domino, Merlin, and Sugar. One way to do this would be to sketch the bands of each dog's DNA fingerprint on the chalkboard so that students can compare the bands. Questions to ask students include: How closely related to each other are the young dogs? *(Domino and Merlin are more closely related than are Domino and Sugar; Merlin and Sugar are more closely related than are Sugar and Domino.)* Are there any bands that all three young dogs have in common? *(All three young dogs have Bella's last band in common.)* Merlin and Sugar have the same mother and father; do they also have the same DNA fingerprints? *(No; Merlin has one of Bella's bands and one of King's bands that Sugar does not have, and Sugar has one of Bella's bands and one of King's bands that Merlin does not have.)*

Bibliography for Teachers

Derr, Mark. "The Making of a Marathon Mutt." *Natural History,* 105 (3): March 1996, p. 34.

Jaroff, Leon. "Keys to the Kingdom." *Time,* 148 (14): Fall 1996, p. 24.

Bibliography for Students

McCaig, "The Dogs that Go to Work, and Play, All Day—for Science." *Smithsonian,* 27 (8): November 1996, p. 126.

Mueller, Larry. "Reproducing Junior." *Outdoor Life,* 165 (2): August 1995, p. 62.

Other Media

Biotechnology
Video
National Geographic Society
Educational Services
P.O. Box 98019
Washington, D.C. 20090-8019
800-368-2728

In addition to the above video, students may find relevant information about genetics and electrophoresis by exploring the Internet. Interested students can search for articles using keywords such as *DNA, DNA fingerprinting, chromosomes, genes,* and *genetic engineering.* Students may also find interesting information about the *Human Genome Project* on the Internet.

Exploration 8

DNA Pawprints

1. Ms. Jean Poole wants to enter her dogs in an upcoming dog show. What does she need to know in order to complete the pedigrees for her dogs?

   _____

   _____

   _____

   _____

2. What does DNA have to do with inherited characteristics? (If you're not sure, check out the CD-ROM articles.)

   _____

   _____

   _____

   _____

   _____

   _____

   _____

3. Explain how DNA fingerprinting works. (Hint: Check out the wall chart.)

   _____

   _____

   _____

   _____

   _____

   _____

   _____

Exploration 8 Worksheet, continued

---

4. Describe the setup on the front table in the lab.

_____

_____

_____

_____

_____

5. Conduct DNA fingerprinting for each young dog, mother, and possible father, and record your results in the table below.

Mother	Young dog	Possible father	Observations of DNA fingerprints
Bella			
Bella			
Bella			
Bella			
Bella			
Bella			
Bella			
Bella			
Bella			

Exploration 8

6. How does Bella's DNA fingerprint compare with the DNA fingerprints of Domino, Merlin, and Sugar?

_____

_____

_____

_____

_____

_____

_____

7. How can you tell which older male fathered each young dog?

_____

_____

_____

_____

_____

_____

_____

_____

_____

_____

8. Look at the material on the lab's back counter. What does it tell you about where DNA is located?

_____

_____

_____

_____

Record your answers in the fax to Ms. Poole.

Name _____ Date _____ Class _____

FAX

To: | Ms. Jean Poole (FAX 512-555-8163)

From: |

Date: |

Subject: | DNA Pawprints

Please indicate which male sired Domino, Merlin, and Sugar.

Domino	○ Duke	○ King	○ Roy
Merlin	○ Duke	○ King	○ Roy
Sugar	○ Duke	○ King	○ Roy

Describe the test that you used to determine which sire matched each of the young dogs.

_____
_____
_____
_____
_____
_____
_____
_____
_____
_____
_____
_____
_____

Name _____ Date _____ Class _____

✿•❈•✿•❈•✿•❈•✿•❈•✿•❈•✿•❈•✿•❈•✿•❈•✿•❈•✿•❈•✿•❈•✿•❈•✿•❈•✿•❈•✿•❈•✿•❈•✿

<u>For Internal Use Only</u>

Please answer the following questions for my laboratory records. Scientists must always keep good records. Dr. Crystal Labcoat

How much genetic information did each young dog inherit from each of its parents? How do you know?

_____

_____

_____

_____

_____

_____

_____

_____

_____

_____

DNA Pawprints

The following articles can also be found by clicking the computer in the CD-ROM laboratory for Exploration 8:

- *Blueprints for Life*
- *DNA Fingerprints*
- *Advances in Genetic Research*

Blueprints for Life

Genes and DNA

When talking about where you get your looks, you may say, "It's in my genes." **Genes** are responsible for the way characteristics are passed on from generation to generation. We inherited our genes from our parents, who inherited their genes from their parents. But what are genes, and how do they determine characteristics such as hair color, eye color, and height? A certain chemical called **DNA** *(deoxyribonucleic acid)* is responsible for the makeup of our genes. DNA is found in almost every one of your cells, and it acts as a set of instructions for how you look and how you inherit or pass on your physical traits. DNA is the blueprint, and you are the finished product.

A molecule of DNA has the structure of a double helix, which resembles a twisted ladder. Each side of the ladder is a separate strand of DNA. Strands of DNA are made up of long chains of smaller units called nucleotides. A **nucleotide** has three parts—a sugar molecule, a phosphate group, and a nitrogen base. The nucleotides differ from one another by the type of nitrogen base present. The four nitrogen bases differ in shape, and their names are adenine (A), thymine (T), guanine (G), and cytosine (C).

Nucleotides are bonded together in a specific way to form the double helix. The sides are formed by bonding the sugar of one nucleotide to the phosphate of the next nucleotide in a continuous chain. Each nitrogen base on one side is bonded to a nitrogen base from the other side to form the rungs of the ladder. The bonded pairs are called **base pairs.** Each base can be paired with only one other base: adenine (A) always pairs with thymine (T), and guanine (G) always pairs with cytosine (C). The order or sequence of the base pairs makes up a gene.

The order of the bases determines what kinds of proteins are made, when they are made, and where they are made. Proteins are responsible for all of the differences in traits among living things. They determine characteristics such as eye color, hair texture, and even how tall you will grow. Although all DNA is made up of the same basic components (a sugar, a phosphate group, and a nitrogen base), not all living things share the same genes. Only in identical twins can you expect to find identical arrangements of nucleotides along a DNA molecule. The millions of arrangements of nucleotides in strands of DNA account for the vast differences in our genetic makeup.

Genes and Chromosomes

Genes occupy specific places on **chromosomes,** which are rod-shaped structures in the nucleus of a cell of a living organism. Chromosomes contain thousands of genes. Humans have 23 pairs of chromosomes in most body cells. As a result of meiosis, sex cells (sperm cells and egg cells) contain only 23 chromosomes apiece—one chromosome from each pair. During sexual reproduction, a sperm cell and an egg cell unite, and the offspring receives 23 chromosomes from the sperm cell (male parent) and 23 chromosomes from the egg cell (female parent) for a total of 46 chromosomes (or 23 pairs). Thus the offspring receives half of its genetic makeup from the mother and the other half from the father.

Passing Traits Along

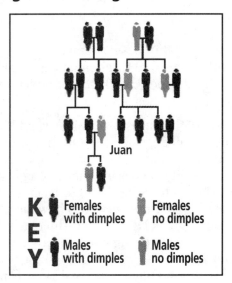

K
E
Y

Females with dimples

Females no dimples

Males with dimples

Males no dimples

What do a dimpled chin, a widow's peak, and green eyes have in common? They are all genetic **traits** that can be passed from one generation to the next. Not all genetic traits show up in consecutive generations, though. Why not? The answer lies in the way genes are expressed.

Remember that genes are located on chromosomes. In a given pair of chromosomes, genes that occupy corresponding locations on each chromosome are called gene pairs, and they code for the same trait. Each member of a gene pair is called an **allele.** Alleles are alternative forms of a gene. In a single gene pair, the alleles may or may not be different. For example, a gene pair that codes for the shape of your hairline might have two different alleles—one that codes for a rounded hairline and one that codes for a widow's peak. Conversely, the gene pair could consist of two alleles for a rounded hairline, or it could consist of two alleles for a widow's peak. Often a trait such as eye or hair color is determined by the interaction of many different gene pairs and alleles.

A person possesses two alleles for each trait. Because humans reproduce sexually, one allele comes from the female parent, and the other comes from the male parent. For example, a mother may carry an allele for a widow's peak while a father may carry an allele for a rounded hairline. The offspring will inherit both alleles of this gene for hairline shape. But what determines which form of the gene will be expressed?

The presence of an allele that codes for a specific trait does not ensure that the trait will be expressed in the individual who carries it. Certain alleles are **dominant;** that is, when the allele is present, it masks the effect of a **recessive** allele. If a person inherits one or more dominant alleles, the dominant trait will be expressed. A person must inherit two recessive alleles for the recessive trait to be expressed. When offspring receive one dominant and one recessive allele, they are carriers of the recessive allele. Although the recessive trait is not generally expressed, the allele can be passed along to future generations. This explains why some traits skip a generation.

DNA Fingerprints

How Can DNA Leave a Fingerprint?

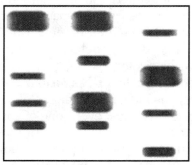

DNA fingerprints can distinguish individuals within a group.

Like fingerprints taken from a person's hand, a DNA fingerprint helps to identify an individual from members of a large population. While taking a thumbprint is a relatively easy procedure, taking a DNA fingerprint requires intensive lab work. First, a sample of DNA must be obtained, usually from blood, hair, or skin. Then the DNA is chemically cut into fragments, and the fragments are sorted by size. Some of these fragments are then tagged with a radioactive probe. When an X ray is taken of the DNA fragments, the fragments appear as a pattern of bands. These patterns allow specialists to determine if two or more DNA samples came from the same person, people who are related, or people who are not related at all.

Cutting DNA Into Fragments

The first step in analyzing DNA involves separating DNA from the rest of the cell parts. The fluid or tissue to be analyzed is transferred to a vial and placed in a centrifuge (a machine that spins material very quickly and causes the DNA to settle at the bottom of the vial). By then heating the DNA, the bonds between the nitrogen base pairs are broken and the strands of DNA come apart. Once the strands have separated, they are chemically cut into fragments. The DNA is then ready for electrophoresis, the process used to separate the DNA fragments.

Electrophoresis

Electrophoresis is a process that uses an electric current to separate the DNA fragments. Two or more samples of DNA are placed in a specially prepared gel tray, and an electric current is passed through the gel tray from one end to the other. The electric current sorts the DNA by size. Because DNA has a negative electrical charge, it will move toward the positive electrode of the electrical source. DNA fragments of different sizes separate because small fragments move toward the positive electrode more quickly than large fragments do. The gel containing the DNA can then be prepared for further examination.

Southern Blot

The Southern Blot is the process that allows us to see the separated fragments of DNA created by the electrophoresis process. The Southern Blot produces a series of bands that allows you to compare one sample of DNA with another. The gel containing the separated fragments of DNA are chemically bathed so that they can be transferred to a nylon membrane. The membrane is

then baked to permanently attach the DNA to it. After another series of chemical baths, fragments of DNA are marked with a radioactive tag, and an X ray of the membrane is taken. The DNA will show up on X-ray film as bands of different widths.

Who Needs to Fingerprint DNA?

DNA fingerprinting has a number of applications. For instance, police departments and forensic specialists can use DNA fingerprinting to determine whether a criminal suspect was present at the scene of a crime. Cells containing the DNA to be tested can come from hair, blood, skin, saliva, or other bodily substances. Missing persons may also be tracked using their DNA profile. If an individual's parentage is unclear, DNA fingerprints may be taken to help determine close family relationships. This is done by examining bands of possibly related parents and offspring. Bands in a child's DNA fingerprint that do not match bands in the mother's DNA fingerprint must match bands in the father's DNA fingerprint. In one particularly interesting case, a community in England is planning to create a database of the DNA profiles of all of its 30 dogs. Apparently, the community is having problems with dogs defecating on the sidewalks. The DNA taken from hair samples will be used to determine which dogs are responsible for the mess.

Exploration 8

Advances in Genetic Research

The Human Genome Project

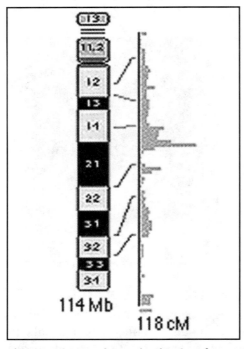

This genetic map shows the density of genes along a given chromosome.

Scientists estimate that humans have over 100,000 different genes. The Human Genome Project is a worldwide effort to analyze all of the genetic material in humans and identify where every gene is located along our 23 pairs of chromosomes. In addition, scientists working on the Human Genome Project plan to learn the base sequence (the order of the bases) for the entire human genome. The **human genome** contains all the genes that make up the master blueprint of a person. The U.S. Human Genome Project began in October 1990. By 1995, over 75 percent of the human genome had been mapped. Researchers plan to use this information to learn more about how genetic mutations play a role in many of today's most common diseases, such as heart disease, diabetes, and birth defects. One day, scientists may be able to treat genetic diseases by correcting errors in the gene itself.

A Gene for Parkinson's Disease

As a result of further genetic research, scientists have been able to pinpoint the location of genes associated with many health disorders. In November 1996, scientists discovered a gene associated with Parkinson's disease. Parkinson's is a progressive disorder that usually strikes adults later in life. Symptoms of this disease include shaky hands, muscular stiffness, and slowness of movement. Mutations (changes in the base-pair order) in the gene will cause classical Parkinson's symptoms. Scientists plan to study this gene in Parkinson's research to help make early diagnoses and to develop new methods of treatment for people currently diagnosed with Parkinson's.

Answer Keys

SciencePlus Interactive Explorations Teacher's Guide

Contents

Answer Key

The Nose Knows

1. Ms. Foushen needs your help sniffing out the solution to a problem. What has she asked you to do?

Ms. Foushen wants to know which of five smelly chemicals would be the best choice for use in a fire alarm at her school for hearing- and sight-impaired students.

(Recommended 10 pts.)

2. Explain the process of diffusion. (Hint: Check out the wall chart in the lab.)

Sample answer: Diffusion is caused by the constant motion of the particles of matter. It is the process in which particles of a substance move from an area of higher concentration to areas of lower concentration until the particles reach a uniform concentration.

(Recommended 20 pts.)

3. What is the difference between diffusion and osmosis? (If you're not sure, check out the CD-ROM articles.)

Sample answer: Osmosis is diffusion through a semipermeable membrane. The membrane permits the passage of some types of particles while preventing the passage of other types. Diffusion takes place among all particles without a semipermeable membrane.

(Recommended 10 pts.)

4. What is the equipment on the front table in Dr. Labcoat's lab designed to do?

Sample answer: It is designed to show how quickly the particles of each smelly chemical diffuse among air particles.

(Recommended 10 pts.)

5. Use the equipment to conduct the necessary tests, and record your data in the table below.

Test tube	Test-tube contents	Time to diffuse (sec.)
A	perfume	15
B	rotten eggs	5
C	garlic	12
D	alcohol	6
E	cinnamon	10

(Recommended 15 pts.)

6. Why are the temperature and pressure kept constant for this experiment? (If you're not sure, check out the CD-ROM articles.)

Sample answer: Temperature and pressure must be kept constant so that the rate of diffusion of each chemical is not affected by different variables. Temperature and pressure both influence the rate of diffusion of a substance. Particles at higher temperatures and pressures diffuse faster because of the increased amount of kinetic energy of the particles.

(Recommended 15 pts.)

7. What does the equipment on the back counter in Dr. Labcoat's lab demonstrate?

Sample answer: This equipment shows how a higher temperature increases the rate of diffusion. The 30°C water allows the ink particles to diffuse faster than the 10°C water does.

(Recommended 10 pts.)

8. How do you smell an odorous chemical? Use the CD-ROM articles to help you explain how your sense of smell works.

Sample answer: After you inhale odor particles that are diffused through the air, some of the odor particles settle on olfactory receptor cells in the nose. These receptor cells send a message along the olfactory nerve to the brain, where the smell is interpreted.

(Recommended 10 pts.)

Record your answers in the fax to Ms. Foushen.

FAX

To: Ms. Dee Foushen (FAX 512-555-7003)

From:

Date:

Subject: The Nose Knows

Which of the five samples do you recommend that I use for the fire alarm?

Alcohol	Cinnamon	Garlic	Perfume	Rotten eggs

(Recommended scoring: 50 points for cinnamon; 30 points for garlic or perfume; and 0 points for alcohol or rotten eggs.)

Please explain why you chose this sample.

The sample that diffused the fastest, the rotten eggs, is highly toxic. The sample that diffused the second fastest, alcohol, is dangerous to use as a fire alarm because it is flammable. Cinnamon is the sample that diffused the third fastest, and it is nontoxic. Therefore, cinnamon would be the best choice to warn the students of fire the most quickly and safely. *(Recommended 25 pts.)*

Explain how odors spread through a room.

Odors spread through a room by a process called diffusion, during which molecules of one substance move from an area of high concentration to an area of lower concentration until a relatively uniform concentration has been reached.

(Recommended 25 pts.)

Sea the Light

1. Ms. Sittie wants to create an underwater hanging lamp. What help does she need from you?

   Ms. Sittie needs to know which ballast disk to add to a waterproof lamp base to make the entire lamp neutrally buoyant underwater where she dives. *(Recommended 10 pts.)*

2. What does *ballast* mean? (If you aren't sure, use the CD-ROM articles to help you.)

   The term *ballast* refers to something heavy that is used to add mass to an object such as a ship. *(Recommended 5 pts.)*

3. What purpose do you think the ballast disks serve in the design of the underwater lamp?

   Depending on which ballast disk is used in combination with the lamp base, the underwater lamp will rise, sink, or be neutrally buoyant underwater. *(Recommended 5 pts.)*

4. Describe how you will use the equipment on the lab's front table to answer Ms. Sittie's questions.

   The equipment can be used to measure the volume and the mass of each of the ballast disks so that the density of each disk can be calculated. Finding out the total density of the lamp base with each individual ballast disk will determine which disk Ms. Sittie should use for her hanging lamp. *(Recommended 10 pts.)*

Answer Key • Exploration 2

5. Use the equipment to conduct all of the necessary tests, and record your data in the first two columns of the table below. Then use your results to calculate the values for the third column.

Ballast disk	Mass (g)	Volume (mL)	Density (g/mL)
A Copper	1344.0	150	8.96
B Aluminum	1469	550	2.67
C Brass	4280.0	500	8.56
D Titanium	1135.0	250	4.54
E Platinum	5363.0	250	21.45
F Zinc	998.0	140	7.13

(Recommended 10 pts.)

6. Calculate the total density of the entire lamp for each individual ballast disk. (If you aren't sure how to calculate the density of an object with multiple parts, examine the CD-ROM articles.)

Ballast disk	Density
A Copper	(659 g + 1344.0 g) ÷ (1500 mL + 150 mL) = 1.214 g/mL
B Aluminum	(659 g + 1469.0 g) ÷ (1500 mL + 550 mL) = 1.038 g/mL
C Brass	(659 g + 4280.0 g) ÷ (1500 mL + 500 mL) = 2.469 g/mL
D Titanium	(659 g + 1135.0 g) ÷ (1500 mL + 250 mL) = 1.025 g/mL
E Platinum	(659 g + 5363.0 g) ÷ (1500 mL + 250 mL) = 3.441 g/mL
F Zinc	(659 g + 998.0 g) ÷ (1500 mL + 140 mL) = 1.010 g/mL

(Recommended 15 pts.)

7. Examine the materials on the back counter of the lab. Use what you see to explain why the different ballast disks have different densities.

The atoms of different elements have different particle arrangements and different densities. Because density is a physical property of an element, each ballast disk has a different density because each disk is made of different elements.

(Recommended 15 pts.)

8. What is buoyant force? (If you're not sure, check out the CD-ROM articles.)

Buoyant force is the upward force exerted by a fluid on a submerged object. The buoyant

force is equal to the weight of the fluid displaced by the object. *(Recommended 15 pts.)*

_____

_____

_____

_____

_____

_____

9. Describe the differences among underwater objects that are positively, negatively, and neutrally buoyant. (Hint: Check out the CD-ROM articles.)

A positively buoyant object will tend to rise underwater because it is less dense than the

surrounding water. A negatively buoyant object will sink because its density is greater than

the density of the surrounding water. A neutrally buoyant object will neither sink nor rise

because its density equals the density of the water. *(Recommended 15 pts.)*

_____

_____

_____

_____

_____

_____

Record your answers in the fax to Ms. Sittie.

Answer Key • Exploration 2

FAX

To: Ms. Diane Sittie (FAX 817-555-4459)

From:

Date:

Subject: Sea the Light

Please complete the following chart:

METAL	MASS	VOLUME	DENSITY
Aluminum	1469.0 g	550 mL	2.67 g/mL
Brass	4280.0 g	500 mL	8.56 g/mL
Copper	1344.0 g	150 mL	8.96 g/mL
Platinum	5363.0 g	250 mL	21.45 g/mL
Titanium	1135.0 g	250 mL	4.54 g/mL
Zinc	998.0 g	140 mL	7.13 g/mL

(Recommended 10 pts.)

What is the density of the waterproof lamp base?

659 g ÷ 1500 mL = 0.439 g/mL   *(Recommended 10 pts.)*

Please indicate your metal selection for the ballast disk here: **Titanium**
(Recommended scoring: 50 points for titanium; 30 points for aluminum or zinc; and 0 points for brass, copper, or platinum.)

Why did you pick this metal?

Sample answer: The total density of the lamp base and the titanium ballast disk is approximately the same as the density of the sea water where Ms. Sittie dives. As a result, the entire lamp will hang in the water at the depth that Ms. Sittie needs. If the lamp base plus the ballast disk were denser than the sea water, then the lamp would sink. If the lamp base plus the ballast disk were less dense than the sea water, then the lamp would float to the surface of the water. *(Recommended 30 pts.)*

Stranger Than Friction

1. Mr. Cline is hard at work on his design for a new amusement park ride. What information is he seeking from you?

   Mr. Cline wants to know which material he should use to construct a slide and to use for

   the bottom of the toboggans for the Camelback Super Slide. He also wants to know what

   size to make the toboggans. *(Recommended 10 pts.)*

2. Describe the equipment Dr. Labcoat has set up on the front lab table and the back counter.

   On the front table is a model of the Camelback Super Slide, three different-sized tobog-

   gans, and a test figure. On the back counter are four blocks, two with a mass of 50 g and

   two with a mass of 100 g. *(Recommended 10 pts.)*

3. What is frictional force, and how does it affect a moving object? (Hint: If you're not sure, check out the CD-ROM articles.)

   Frictional force is the force that opposes motion between two surfaces that are touching.

   Frictional force works to keep an object from moving or slows an object down that is

   already in motion. The magnitude of frictional force depends on the coefficient of friction

   and on the normal force. *(Recommended 10 pts.)*

4. What does the wall chart in the lab show about normal force and the coefficient of friction?

   The diagram shows how a greater normal force (the force pressing two objects together)

   can result in a greater frictional force. Therefore, a greater amount of force is necessary to

   move heavier objects. It also shows how changing the coefficient of friction between two

   surfaces can reduce friction and the effects of frictional force. *(Recommended 20 pts.)*

Answer Key • Exploration 3

5. Use the force meter on the back lab counter to find the force required to pull each block. Record your results below.

It takes 0.25 N of force to pull both 50 g blocks at the same speed. It takes 0.50 N of force

to pull both 100 g blocks at the same speed. *(Recommended 15 pts.)*

6. Does the amount of surface area touching the block affect the force required to pull it? Why or why not?

Both 50 g blocks require the same amount of force to move across the counter at the same

speed, even though one block is on its side and the other is on its edge. The same is true

for both 100 g blocks. This shows that the force of friction is not determined by the area of

contact between two surfaces. *(Recommended 15 pts.)*

7. Conduct the necessary tests with the prototype for the Camelback Super Slide, and record your results in the table below. (Hint: It may not be necessary to try every possible combination.)

Material component for slide	Material component for toboggan	Toboggan size (cm)	Observations
teflon	teflon	any	Toboggan loses contact with the slide at the hump. Figure flies off the slide.
teflon	stainless steel	any	Toboggan slips on the hump and ejects figure. Both land at bottom of slide.
teflon	plastic	any	Toboggan gets stuck and slides back and forth in dip before hump.
stainless steel	teflon	any	Toboggan slips on the hump and ejects figure. Both land at bottom of slide.
stainless steel	stainless steel	any	Toboggan stays in contact with slide all the way down, giving the best ride.
stainless steel	plastic	any	Toboggan gets stuck and slides back and forth in dip before hump.
plastic	teflon	any	Toboggan gets stuck and slides back and forth in dip before hump.
plastic	stainless steel	any	Toboggan gets stuck and slides back and forth in dip before hump.
plastic	plastic	any	Toboggan gets stuck and slides back and forth in dip before hump.

(Recommended 20 pts.)

Record your answers in the fax to Mr. Cline.

FAX

To: Mr. Norm N. Cline (FAX 281-555-5276)

From:

Date:

Subject: Stranger Than Friction

What material do you recommend for the construction of the slide?

Stainless steel

What material do you recommend for the construction of the toboggan?

Stainless steel

What is your recommendation regarding toboggan size?

100 cm
120 cm
140 cm
any of the above

(Recommended scoring: 50 points for stainless steel for both the slide and the toboggans and "any of the above" selected; 30 points for one or two incorrect materials chosen and "any of the above" selected; 30 points for stainless steel for both the slide and the toboggans and "any of the above" not selected; 0 points for one or two incorrect materials and "any of the above" not selected.)

What effect does the size of the toboggan have on the performance of the Camelback Super Slide? Explain.

The size of the toboggan has no effect on the performance of the ride. The friction between two given surfaces is the same for a given normal force no matter how much area is in contact between the two surfaces. *(Recommended 50 pts.)*

Latitude Attitude

1. Capt. Corey O. Lease needs to get supplies to a group of scientists on the Mertz Glacier. What has she asked you to do for her?

Capt. Lease has asked Dr. Labcoat to verify that her flight direction and average speed

calculations are appropriate for a trip from Melbourne to the Mertz Glacier.

(Recommended 10 pts.)

2. In relation to Melbourne, where is the Mertz Glacier located?

The Mertz Glacier is due south of Melbourne. *(Recommended 10 pts.)*

3. How does the speed of the Earth's surface at the equator compare with the speed of the Earth's surface at the poles? (If you're not sure, check out the CD-ROM articles.)

Points on the Earth's surface at the equator have greater velocities (about 1670 km/hr) than

do points at the poles (effectively 0 km/hr). *(Recommended 15 pts.)*

4. How will you use the Accu-Flight Simulator to help Capt. Lease?

Sample answer: I will use the Accu-Flight Simulator to examine the results of different com-

binations of flight direction and airspeed. After selecting a certain flight direction and air-

speed, I will look at the display screen to see where Capt. Lease's plane would end up.

When I find the combination of flight direction and airspeed that takes the plane directly to

the Mertz Glacier while maintaining fuel efficiency, I will recommend this flight plan to

Capt. Lease. *(Recommended 15 pts.)*

5. Use the Accu-Flight Simulator to determine the results of using different combinations of flight direction and average airspeed. Record your results in the table below.

Flight direction	Airspeed (km/hr)	Final location of the plane
130°	725	east of the Mertz Glacier, off the screen
140°	725	east of the Mertz Glacier, off the screen
150°	725	northeast of the Mertz Glacier, in the ocean
180°	725	northeast of the Mertz Glacier, in the ocean
210°	725	on Antarctica, but east of the Mertz Glacier
220°	725	on Antarctica, slightly east of the Mertz Glacier
230°	725	on Mertz Glacier
130°	850	east of the Mertz Glacier, off the screen
140°	850	east of the Mertz Glacier, off the screen
150°	850	northeast of the Mertz Glacier, in the ocean
180°	850	on Antarctica, but east of the Mertz Glacier
210°	850	on Antarctica, but east of the Mertz Glacier
220°	850	on Mertz Glacier
230°	850	on Antarctica, slightly west of the Mertz Glacier

(Recommended 15 pts.)

6. What does the equipment on the lab's back counter demonstrate?

The equipment demonstrates why the Coriolis effect occurs. The turntable rotates clockwise about one-third of a revolution. As a result, the pencil line curves to the left. This is similar to what happens to Capt. Lease's plane as it travels toward Antarctica.

(Recommended 15 pts.)

7. Describe how moving objects are affected by the Coriolis effect. (If you're not sure, check out the CD-ROM articles.)

The Coriolis effect influences the paths of moving objects such as airplanes, winds, and weather systems. In the Northern Hemisphere, objects moving north or south seem to veer to their right (clockwise). In the Southern Hemisphere, the curvature of the path of objects moving north or south is to their left (counterclockwise). *(Recommended 20 pts.)*

Record your answers in the fax to Capt. Lease.

FAX

To:	Capt. Corey O. Lease (FAX 011-612-555-7996)
From:	
Date:	
Subject:	Latitude Attitude

In which compass direction should I aim my plane?

180° (due south)	130°	140°	150°	210°	220°	230°

How fast should I fly the plane?

725 km/hr	850 km/hr

(Recommended scoring: 50 points for 230° and 725 km/hr; 35 points for 220° and 850 km/hr; and 0 points for anything else.)

Please explain why it is necessary for me to fly the plane in the recommended direction in order to reach the Mertz Glacier.

Sample answer: Points on the Earth's surface at higher latitudes have less surface velocity than those at latitudes closer to the equator. As a result, the paths of objects that move above the Earth's surface appear curved. This phenomenon is called the Coriolis effect. For example, an object moving south in the Southern Hemisphere, such as Capt. Lease's plane, is deflected counterclockwise, or east, in relation to the ground. Because of the Coriolis effect, Capt. Lease must aim her plane slightly southwest to reach her destination. The exact direction in which she must fly depends on how fast the plane is traveling. For example, if Capt. Lease flies her plane at an average speed of 850 km/hr, the angle southwest at which she flies the plane may be less than it would be if she flies her plane at 725 km/hr.

(Recommended 50 pts.)

Exploration 5
Worksheet

Tunnel Vision

1. Seymore Rhodes wants to make his mountain-bike ride to school a safer one. What advice does he need from you?

   Seymore wants to know how to brighten the lights in a circuit for his bicycle helmet. He

   also wants to make sure that the circuit is lightweight. *(Recommended 10 pts.)*

2. What are the necessary components of an electric circuit? (If you're not sure, check out the CD-ROM articles.)

   The components of an electric circuit include a source of electricity, such as a battery; an

   output device, such as a light bulb; and wires. Many circuits also include switches, which

   allow the circuit to be turned on and off. *(Recommended 10 pts.)*

3. Look at the diagram that Seymore drew of his circuit. What kind of circuit is it?

   Seymore's circuit is a series circuit. *(Recommended 5 pts.)*

4. How is a series circuit different from a parallel circuit? (Hint: Check out the equipment on the lab's back counter.)

   In a series circuit, there is only one pathway for the current to follow. If there is a break in

   the circuit, the flow of current is disrupted and none of the bulbs light. In a parallel circuit,

   current is divided into two or more branches. If a break occurs in one branch, current can

   continue to flow through other branches and the bulbs along those branches stay lit. On the

   back counter, the bulbs in the series circuit are not as bright as those in the parallel circuit.

   This is because the bulbs connected in series share the same current, while each light bulb

   connected in parallel receives the full amount of current provided by the battery.

   (Recommended 15 pts.)

Answer Key • Exploration 5

5. Examine the equipment on the lab's front table. How will you use the switches and the batteries to help you answer Seymore's questions?

Sample answer: I will use the switches to create various circuits. By closing certain switches, I can observe how different circuits affect the operation and brightness of the three light bulbs. I will use the batteries to determine how greater amounts of voltage affect the brightness of the bulbs. *(Recommended 10 pts.)*

6. Create all of the necessary circuits you need to answer Seymore's questions. In the table below, record the switches you close, the number of batteries you use, and the type of circuit you create. Use the fourth column to record your observations about the light bulbs.

Please note: The following are sample results. Many more combinations are possible.

Switches closed	Number of batteries	Type of circuit	Effect on light bulbs
1, 3, 6	1	series	A and B are somewhat bright.
1, 2, 4, 6	1	parallel	A and B are very bright.
1, 3, 5, 8	1	series	A, B, and C are dim. (This is Seymore's circuit.)
1, 3, 6, 7, 8	1	series-parallel	A and B are somewhat bright; C is very bright.
1, 2, 4, 5, 8	1	series-parallel	A is very bright; B and C are somewhat bright.
1, 2, 4, 6, 7, 8	1	parallel	A, B, and C are very bright. (This is the best circuit.)
1, 3, 5, 8	2	series	A, B, and C are somewhat bright.
1, 2, 4, 6	2	parallel	A and B burn out.
1, 2, 7, 8	2	parallel	A and C burn out.
1, 2, 4, 5, 8	2	series-parallel	A burns out; B and C are very bright.
1, 2, 4, 6, 7, 8	2	parallel	A, B, and C burn out.

(Recommended 20 pts.)

7. What effect does the number of batteries have on your circuits?

The number of batteries affects the brightness of the bulbs because the batteries determine the amount of current that can flow through the circuits created in this Exploration. For example, in a series circuit consisting of three bulbs, adding a second battery made the bulbs brighter. In a parallel circuit consisting of three bulbs, adding a second battery caused the bulbs to burn out. Adding a second battery also increases the total weight of the circuit.   *(Recommended 10 pts.)*

8. What happens when too much current flows through a circuit?

When too much current flows through a circuit, the circuit can fail. For example, if too much current is introduced into a circuit involving a light bulb, the bulb can burn out.
                                                                                       (Recommended 10 pts.)

9. What is the relationship between current, voltage, and resistance? (If you're not sure, check out the CD-ROM articles.)

Ohm's law describes the relationship between current, voltage, and resistance. For a given resistance, current and voltage are directly proportional according to the equation $I = E/R$. That means that, for a given resistance, increasing the voltage increases the current.
                                                                                       (Recommended 10 pts.)

Record your answers in the fax to Seymore Rhodes.

FAX

To: Seymore Rhodes (FAX 406-555-7209)

From:

Date:

Subject: Tunnel Vision

How many batteries should I use? | 1 | 2 |

Please describe the best way to wire the lights for my helmet.

The best way to wire the lights for the bicycle helmet is in parallel with one battery.

With this combination, the bulbs are very bright and the circuit is still lightweight.

(Recommended 25 pts.)

For Internal Use Only

Please answer the following questions for my laboratory records. Scientists must always keep good records. Dr. Crystal Labcoat

What is the best type of circuit for Seymore's lights?

| SERIES | SERIES-PARALLEL | PARALLEL |

(Recommended scoring: 50 points for one battery and parallel; 30 points for two batteries and series; and 0 points for any other answers.)

Describe the changes that you made to Seymore's original circuit.

Seymore's original circuit is a series circuit. Because the three light bulbs are connected in series, they share the same current. As a result, the current produced by the single battery is only strong enough to dimly light all three bulbs. In order to improve Seymore's circuit, it is necessary to connect the bulbs in parallel. This means that each light bulb will receive the full strength of the current provided by the battery, and the bulbs will be very bright. *(Recommended 25 pts.)*

Answer Key • Exploration 5

Sound Bite!

1. Mr. Lintz is worried about the aggressive behavior of his guinea pigs. Describe his problem and what he has asked you to do.

The guinea pigs in Mr. Lintz's pet store have been biting the customers. Mr. Lintz suspects that the loud humming noise from the newly installed freezers next door is driving his guinea pigs to violence. He wants to know how to use active noise control to eliminate this noise pollution. *(Recommended 10 pts.)*

2. Examine the equipment on the back counter of the lab, and then describe what sound waves are.

As demonstrated by the equipment, air particles spread away from a vibrating sound source, such as the drum, and push against air particles in front of them. The particles continue to crowd together and spread apart in succession as the sound travels away from its source, and the compressions and rarefactions of longitudinal waves of sound result. *(Recommended 15 pts.)*

3. What purpose does an oscilloscope serve? (Hint: Check out the wall chart.)

An oscilloscope is an instrument that converts sound waves (longitudinal waves) to transverse waves that appear on a screen. This allows us to "see" sound waves. *(Recommended 10 pts.)*

4. What is destructive interference? (If you're not sure, check out the CD-ROM articles.)

Destructive interference occurs when the crest of one wave meets the trough of another wave. This causes a decrease in amplitude. If a crest and trough of equal amplitude and frequency meet, they cancel each other out. *(Recommended 15 pts.)*

5. Conduct your experiment using the lab equipment. In the table below, record the frequency, amplitude, and phase you selected for the generated sound. Use the fourth column to describe the wave you see in the center oscilloscope screen when the generated sound is combined with the offending sound.

Please note: The following answers are sample observations. Many more combinations are possible.

Frequency (Hz)	Amplitude	Phase	Results
175	low	1	A wave with varying amplitude. When the amplitude is at its greatest, it is greater than the amplitude of both the offending sound and the generated sound.
225	low	1	A wave with constant amplitude. The amplitude is twice that of either the offending sound or the generated sound.
275	low	1	A wave with varying amplitude. When the amplitude is at its greatest, it is greater than the amplitude of both the offending sound and the generated sound.
325	low	1	A wave with varying amplitude. When the amplitude is at its greatest, it is greater than the amplitude of both the offending sound and the generated sound.
175	med.	1	A wave with varying amplitude. When the amplitude is at its greatest, it is greater than the amplitude of both the offending sound and the generated sound.
225	med.	1	A wave with constant amplitude. The amplitude is more than twice that of the offending sound.
325	high	1	A wave with varying amplitude. When the amplitude is at its greatest, it is greater than the amplitude of both the offending sound and the generated sound.
175	low	2	A wave with varying amplitude. When the amplitude is at its greatest, it is greater than the amplitude of both the offending sound and the generated sound.
225	low	2	A flat line appears on the screen. The crests and troughs of the waves of the offending sound and the generated sound cancel out.
275	med.	2	A wave with varying amplitude. When the amplitude is at its greatest, it is greater than the amplitude of both the offending sound and the generated sound.
225	high	2	A wave with constant amplitude. The amplitude is more than twice that of the offending sound but less than the generated sound.

(Recommended 25 pts.)

Answer Key • Exploration 6

6. Based on your results, what are the offending sound's frequency, amplitude, and phase?

The offending sound is in phase 1 and has a frequency of 225 Hz and a low amplitude.

(Recommended 10 pts.)

7. What does *phase* refer to? (If you're not sure, check out the CD-ROM articles.)

***Phase* is a term used to describe the positions of the crests and troughs of two individual**

waves relative to each other. *(Recommended 15 pts.)*

Record your answers in the fax to Mr. Lintz.

FAX

To: Mr. Cy Lintz (FAX 707-555-8988)

From:

Date:

Subject: Sound Bite!

Which settings eliminate Mr. Lintz's noise pollution?

Frequency (hertz)	175	225	275	325
Amplitude	low	medium	high	
Phase	1	2		

(Recommended scoring: 50 points for all three correct answers; 30 points for one or two correct answers; and 0 points for no correct answers.)

For Internal Use Only

Please answer the following questions for my laboratory records. Scientists must always keep good records. Dr. Crystal Labcoat

Explain how sound travels.

Sample answer: Sound waves are longitudinal waves that travel through a medium, such as air. As a source of sound vibrates, it alternately pushes the air molecules together, forming compressions, and spreads them apart, forming rarefactions. A series of compressions and rarefactions form and spread out from the source of the sound. *(Recommended 25 pts.)*

Explain how active noise control is used to eliminate noise pollution.

Sample answer: Active noise control uses destructive interference to reduce a sound's amplitude. A sound of the same frequency as the noise pollution is used to cancel out the offending noise. In some cases, a speaker is set up to produce sound waves that are the mirror image of the noise pollution. When the sound waves from the speaker meet the sound waves from the source of noise, the waves cancel each other out. *(Recommended 25 pts.)*

In the Spotlight

1. Ms. Kones is a little in the dark about the lighting design for her first stage production. What questions do you need to answer for her?

Ms. Kones wants to know how many different colors of light she can create using her filters,

what those colors are, and what filter combinations are required to make those colors. She

also wants to know why blue light added to yellow light results in white light.

(Recommended 10 pts.)

2. Describe the equipment that Dr. Labcoat has set up on the front table.

There is a white screen, two spotlights, and 7 filter selections (including no filter) installed

in each spotlight. *(Recommended 5 pts.)*

3. How do colored filters work to create different colors of light? (If you aren't sure, check out the CD-ROM articles.)

A filter is a colored lens that absorbs certain wavelengths of light and transmits others.

When white light passes through a filter, the filter absorbs all the colors of light except the

color of light that matches the color of the filter. A red filter transmits only red light and

absorbs every other color of light. Because the filter absorbs some colors of light, using

filters is a kind of subtractive color formation. Filters can also be used in additive color

formation. By mixing two different colors of filtered light, new colors can be produced.

(Recommended 15 pts.)

4. As you try different combinations of filters, record your results, including the color equations for each color, in the table below.

Filter 1	Filter 2	Resulting color	Color equation
no filter	no filter	white	no filter + no filter = white
no filter	red	red	no filter + red = red
no filter	cyan	cyan	no filter + cyan = cyan
no filter	yellow	yellow	no filter + yellow = yellow
no filter	blue	blue	no filter + blue = blue
no filter	green	green	no filter + green = green
no filter	magenta	magenta	no filter + magenta = magenta
red	red	red	red + red = red
red	cyan	white	red + cyan = white
red	yellow	burnt orange	red + yellow = burnt orange
red	blue	magenta	red + blue = magenta
red	green	yellow	red + green = yellow
red	magenta	raspberry	red + magenta = raspberry
cyan	cyan	cyan	cyan + cyan = cyan
cyan	yellow	green	cyan + yellow = green
cyan	blue	medium blue	cyan + blue = medium blue
cyan	green	pale green	cyan + green = pale green
cyan	magenta	blue	cyan + magenta = blue
yellow	yellow	yellow	yellow + yellow = yellow
yellow	blue	white	yellow + blue = white
yellow	green	lemon lime	yellow + green = lemon lime
yellow	magenta	red	yellow + magenta = red
blue	blue	blue	blue + blue = blue
blue	green	cyan	blue + green = cyan
blue	magenta	vivid violet	blue + magenta = vivid violet
green	green	green	green + green = green
green	magenta	white	green + magenta = white

(Recommended 20 pts.)

5. What are the primary colors of light?

The primary colors of light are red, blue, and green. *(Recommended 10 pts.)*

6. What are the secondary colors of light?

The secondary colors of light are yellow, cyan, and magenta.

(Recommended 10 pts.)

7. What happens when you mix two primary colors of light?

Mixing two primary colors of light results in a secondary color of light.

(Recommended 10 pts.)

8. What happens when you mix two secondary colors of light?

Mixing two secondary colors of light results in a primary color of light.

(Recommended 10 pts.)

9. What differences do you notice between mixing the colors of paint on the back counter and mixing colors of light? (If you aren't sure, check out the CD-ROM articles.)

Sample answer: When you mix colors of paint, more colors are absorbed than are reflected, so the combinations move toward "blackness" instead of "whiteness." As a result, red + blue + yellow = "black." Mixing colors of paint is subtractive color formation, unlike mixing colors of light, which is additive color formation. *(Recommended 10 pts.)*

Record your answers in the fax to Ms. Kones.

FAX

To: Ms. Iris Kones (FAX 409-555-2017)

From:

Date:

Subject: In the Spotlight

How many different colors of light (including white light) can be produced using Ms. Kones's two spotlights and twelve filters? **13** *(Recommended scoring: 50 points for 13 colors; 30 points for 10, 11, or 12 colors; and 0 points for 9 or fewer colors.)*

Write the color equation for each color that you produced.

White: no filter + no filter = white; yellow + blue = white; green + magenta = white; red + cyan= white

Red: red + no filter = red; red + red = red; yellow + magenta = red

Green: green + no filter = green; green + green = green; cyan + yellow = green

Blue: blue + no filter = blue; blue + blue = blue; cyan + magenta = blue

Magenta: magenta + no filter = magenta; magenta + magenta = magenta; red + blue = magenta

Yellow: yellow + no filter = yellow; yellow + yellow = yellow; red + green = yellow

Cyan: cyan + no filter = cyan; cyan + cyan = cyan; blue + green = cyan

Burnt Orange: red + yellow = burnt orange

Raspberry: red + magenta = raspberry

Medium Blue: cyan + blue = medium blue

Pale Green: cyan + green = pale green

Lemon Lime: yellow + green = lemon lime

Vivid Violet: blue + magenta = vivid violet *(Recommended 15 pts.)*

Describe how colored filters produce different colors of light.

Sample answer: A filter is a colored lens that absorbs certain wavelengths of light and transmits others. For example, when white light is passed through a red filter, red light is transmitted and all other colors of light are absorbed. When two or more beams of filtered light overlap on a white screen, they combine to form a new color of light. *(Recommended 15 pts.)*

Please explain the difference between mixing colors of paint and mixing colors of light.

Sample answer: Mixing colors of paint is called subtractive color formation because it tends to subtract colors of light from white light. For example, if you mix two colors of paint together, the mixture absorbs more colors of light than either paint would absorb separately. When you mix various colors of light, you are adding different wavelengths of light together and "building" white light. This is called additive color formation. *(Recommended 20 pts.)*

DNA Pawprints

1. Ms. Jean Poole wants to enter her dogs in an upcoming dog show. What does she need to know in order to complete the pedigrees for her dogs?

   Ms. Poole needs to know which of her older male dogs fathered each of her three younger

   dogs. She also wants to know a little bit about the test used to determine who sired which

   dog. *(Recommended 5 pts.)*

2. What does DNA have to do with inherited characteristics? (If you're not sure, check out the CD-ROM articles.)

   Sample answer: DNA is the chemical responsible for the makeup of our genes. DNA is

   found in almost every one of our cells, and it acts as a set of instructions for how we look.

   We inherit our genes from our parents, who inherited their genes from their parents. The

   millions of arrangements of nucleotides in strands of DNA account for the vast differences

   in our individual genetic makeup. *(Recommended 10 pts.)*

3. Explain how DNA fingerprinting works. (Hint: Check out the wall chart.)

   DNA fingerprinting is a process that allows you to compare different samples of DNA. An

   electric current sorts fragments of DNA by size in a gel tray. The fragments are transferred

   to a thin film that is bathed in a radioactive solution. The radioactive bath marks certain

   pieces of DNA, which then show up as bands in an X ray. *(Recommended 10 pts.)*

4. Describe the setup on the front table in the lab.

There are DNA samples from Bella, Domino, Merlin, Sugar, Duke, King, and Roy; sterile tips for a micropipet; a gel tray; an electrophoresis chamber; a gel processor; and an X-ray developer.   *(Recommended 10 pts.)*

5. Conduct DNA fingerprinting for each young dog, mother, and possible father, and record your results in the table below.

Mother	Young dog	Possible father	Observations of DNA fingerprints
Bella	Domino	Duke	None of Domino's bands match Duke's bands. Two of Domino's bands match Bella's bands.
Bella	Domino	King	None of Domino's bands match King's bands. Two of Domino's bands match Bella's bands.
Bella	Domino	Roy	Two of Domino's bands match Roy's bands. Two of Domino's bands match Bella's bands.
Bella	Merlin	Duke	None of Merlin's bands match Duke's bands. Two of Merlin's bands match Bella's bands.
Bella	Merlin	King	Two of Merlin's bands match King's bands. Two of Merlin's bands match Bella's bands.
Bella	Merlin	Roy	None of Merlin's bands match Roy's bands. Two of Merlin's bands match Bella's bands.
Bella	Sugar	Duke	None of Sugar's bands match Duke's bands. Two of Sugar's bands match Bella's bands.
Bella	Sugar	King	Two of Sugar's bands match King's bands. Two of Sugar's bands match Bella's bands.
Bella	Sugar	Roy	None of Sugar's bands match Roy's bands. Two of Sugar's bands match Bella's bands.

(Recommended 30 pts.)

6. How does Bella's DNA fingerprint compare with the DNA fingerprints of Domino, Merlin, and Sugar?

Because Bella is the mother of the three dogs, each young dog gets half of its DNA from

Bella. As a result, two bands on each young dog's DNA fingerprint will match two bands on

Bella's DNA fingerprint. *(Recommended 15 pts.)*

7. How can you tell which older male fathered each young dog?

In order for an older male and a younger male dog to be related, they must have some of

the same DNA. Half the young dog's DNA must come from the mother, and the other half

must come from the father. If half the bands of the DNA fingerprints of a young dog and an

older male dog match, then they are related. *(Recommended 15 pts.)*

8. Look at the material on the lab's back counter. What does it tell you about where DNA is located?

DNA is found in the nucleus of almost every cell in all living organisms.

(Recommended 5 pts.)

Record your answers in the fax to Ms. Poole.

FAX

To: Ms. Jean Poole (FAX 512-555-8163)

From:

Date:

Subject: DNA Pawprints

Please indicate which male sired Domino, Merlin, and Sugar.

Domino ◯ Duke ◯ King 🔵 Roy
Merlin ◯ Duke 🔵 King ◯ Roy
Sugar ◯ Duke 🔵 King ◯ Roy

(Recommended scoring: 50 points for all three correct matches; 30 points for one or two correct matches; and 0 points for no correct matches.)

Describe the test that you used to determine which sire matched each of the young dogs.

The test used to determine which sire matched each of the young dogs is called DNA fingerprinting. DNA samples from each dog were sliced into fragments and separated according to size with an electric current. The samples were then marked with a radioactive tag, and an X ray was taken of them to reveal the DNA fingerprint. Each young dog's DNA fingerprint could then be compared with the DNA finger-prints of Bella and each possible father to determine the heredity of each young dog. *(Recommended 30 pts.)*

For Internal Use Only

Please answer the following questions for my laboratory records. Scientists must always keep good records. Dr. Crystal Labcoat

How much genetic information did each young dog inherit from each of its parents? How do you know?

Because mammals reproduce sexually, each young dog receives half of its chromo-somes from its mother and the other half from its father. This is also evident from the DNA fingerprints of the young dogs. Each young dog shares half of its bands with its mother and the other half with its father. *(Recommended 20 pts.)*